The Animal Kingdom: A Very Short Introduction

VERY SHORT INTRODUCTIONS are for anyone wanting a stimulating and accessible way into a new subject. They are written by experts, and have been translated into more than 45 different languages.

The series began in 1995, and now covers a wide variety of topics in every discipline. The VSI library now contains over 500 volumes—a Very Short Introduction to everything from Psychology and Philosophy of Science to American History and Relativity—and continues to grow in every subject area.

Titles in the series include the following:

Peter Holland

THE ANIMAL KINGDOM

A Very Short Introduction

OXFORD
UNIVERSITY PRESS

OXFORD
UNIVERSITY PRESS

Great Clarendon Street, Oxford ox2 6DP

Oxford University Press is a department of the University of Oxford.
It furthers the University's objective of excellence in research, scholarship,
and education by publishing worldwide in

Oxford New York

Auckland Cape Town Dar es Salaam Hong Kong Karachi
Kuala Lumpur Madrid Melbourne Mexico City Nairobi
New Delhi Shanghai Taipei Toronto

With offices in

Argentina Austria Brazil Chile Czech Republic France Greece
Guatemala Hungary Italy Japan Poland Portugal Singapore
South Korea Switzerland Thailand Turkey Ukraine Vietnam

Oxford is a registered trade mark of Oxford University Press
in the UK and in certain other countries

Published in the United States
by Oxford University Press Inc., New York

British Library Cataloguing in Publication Data

Data available

Library of Congress Cataloging in Publication Data

Data available

Typeset by SPI Publisher Services, Pondicherry, India
Printed in Great Britain on acid-free paper by
Ashford Colour Press Ltd., Gosport, Hampshire.

ISBN 978–0–19–959321–7

11

Contents

Acknowledgements

The structure and content of this book owes much to past and present students at the University of Oxford and the University of Reading. Teaching a course in animal diversity to demanding and critical students has forced me to think carefully about the subject; student feedback has also helped highlight key issues. I acknowledge assistance from Merton College, Oxford, and members of the Department of Zoology, University of Oxford, notably Simon Ellis and Penny Schenk. I also thank Max Telford, Claus Nielsen, Bill McGinnis, Stu West, Theresa Burt de Perera, Tobias Uller, Sally Leys, and Per Ahlberg for comments on various sections, and Tatiana Solovieva for drawing the diagrams.

List of illustrations

Chapter 1
What is an animal?

I am the very model of a modern Major-General,
I've information vegetable, animal, and mineral.
Gilbert and Sullivan, *The Pirates of Penzance* (1879)

To build an animal

In our everyday experiences, it is simple to decide which living
things are animals and which are not. Walking through a town we
may encounter cats, dogs, birds, snails, and butterflies, and we
recognize all of these as animals. We should also include humans
in our list. In contrast, we would have no doubt that the trees,
grasses, flowers, and fungi we encounter are not animals, even
though they too are living organisms. The problem of defining or
recognizing an 'animal' starts to arise when we consider some of
the more unusual living organisms, many of which are
microscopic. It is helpful, therefore, to search for precise criteria
for answering the question 'What is an animal?'

One feature shared by all animals is that they are 'multicellular'.
That is, their bodies are made of many specialized cells. By this
criterion, single-celled organisms such as the familiar *Amoeba* are
not considered to be animals, contrary to the views of a century
ago. Indeed, many biologists now carefully avoid the term

'protozoa' for organisms such as *Amoeba*, since by definition an organism cannot be both 'proto' (meaning 'first' and implying one cell) and 'zoa' (implying animal).

Having a body built from many cells is a necessary criterion, but it is not sufficient on its own. The same property is also found in plants, fungi, and some other organisms such as slime moulds, none of which are animals. A second important character of animals is that they get the energy necessary for life by eating other organisms, or parts of organisms, either dead or alive. This is in contrast to green plants, which can harness the Sun's energy using the chemical reaction photosynthesis taking place inside chloroplasts. There are plants that supplement photosynthesis with feeding (for example, the Venus fly trap) and animals with living green algae inside them (for example, corals and green hydra), but these do not blur the essential distinction too much.

Another feature that is often cited is the ability of animals to move and to sense their environment. This criterion holds up well for animals, but we need to remember that many plants have parts that can move, while cellular slime moulds (which are not animals) can form a slowly migrating slug-like structure.

The generation of sperm and egg cells of quite different sizes is another property typical of animals, and one with profound implications for the evolution of animal behaviour, but it is not a character that is readily observed. Perhaps the most consistent structural character is to be found when the cells of adult animals are examined closely. Although animals have many different types of cells, there is one type that has influenced the entire biology of animals and the evolution of the Animal Kingdom. The cell in question is the epithelial cell. These are brick-shaped or column-shaped cells, lacking the rigid cell walls found in plants. Epithelial cells are arranged into flexible sheets with specialized proteins holding neighbouring cells together and other proteins sealing the gaps between cells to make a waterproof layer. Sheets of cells are also found in plants, but their structure is quite different, being less flexible and more permeable.

The epithelial cell sheet of animals is remarkable for both functional and structural reasons. Epithelia can control the chemical compositions of liquids either side of the sheet, allowing animals to create fluid-filled spaces for purposes as diverse as bodily support or the concentration of waste products. Fluid-filled spaces were among the earliest skeletal structures of animals and were a factor permitting an increase in size during evolution, together with energy-efficient locomotion.

In addition, epithelial cell sheets are strong but flexible, supported by a thick layer of proteins such as collagen, allowing precise folding movements to occur. This is particularly important during embryonic development in animals, when folding movements are used to generate the structure of the animal body, rather like miniature origami. In fact, it is quite simple to mimic the earliest stages of animal development using sheets of paper. Although the details differ between species, typical animal development passes through a stage comprised of a ball of epithelial cells (the blastula), which itself was formed by a series of cell divisions from a single cell – the fertilized egg. In most animal embryos, the ball of cells then folds inwards at one point or along a groove, moving some of the cells inside. This event, which forms a tube destined to become the gut, is the crucial step called gastrulation. The indented ball is called a gastrula. Further folding events can occur to form fluid-filled supporting structures, muscle blocks, and – in vertebrates such as ourselves – even the spinal cord and brain. In short, cell sheets build animals.

All these characters are criteria by which we recognize animals, and they give insight into the basic biology of animals. But they do not comprise the most precise definition of an animal. In taxonomy, the classification of living organisms, names are given to branches – large or small – on evolutionary trees. An essential rule is that real or 'natural' groupings must encompass sets of organisms that have a shared evolutionary ancestor. This means that the term 'animal' must refer to a group of related species. The word cannot be applied to living organisms from elsewhere in the evolutionary tree, even if they possess some animal-like characters. Likewise, we would still use the term 'animal' for species that had lost some of the normal animal characters that were present in their ancestors. For example, some animals have lost distinct sperm and egg cells in evolution, while others are not clearly multicellular in every part of their life cycle, but since they share an ancestor with other animals, they are defined as animals. The animals, therefore, are a natural group (or clade) descended from a shared common ancestor. This clade is called the Animal Kingdom, or Metazoa.

The origin of animals

From what did the long-extinct ancestor of all animals evolve? This sounds a difficult problem to solve, since the ancestor in question has been extinct for perhaps 600 million years, was certainly microscopic, and has left no fossil record. Surprisingly, the answer is known with considerable confidence. Furthermore, it was first suggested over 140 years ago. In 1866, the American microscopist, philosopher, and biologist Henry James Clark noted that the feeding cells of sponges, which are certainly animals, look quite similar to a little-known group of single-celled aquatic organisms known then as the 'infusoria flagellata'. Today, we call these microscopic organisms the choanoflagellates (collared flagellates) and DNA sequence comparisons confirm that they are indeed the closest relatives of all animals. Choanoflagellates and the feeding cells of sponges each have a circle, or 'collar', of fine tentacles at one end, such that they resemble miniature badminton shuttlecocks, plus a single long flagellum (a moveable whip-like structure) emerging from the middle of the collar. In choanoflagellates, the wafting or beating of the flagellum sets up water currents to bring food particles towards the cell, where they

1. A choanoflagellate, *Monosiga brevicollis*, feeding on bacteria

are trapped by the collar. The feeding cells of sponges operate in a different way, but they still use the flagellum to create a water current. The most recent ancestor of all animals, therefore, was probably a microscopic ball of cells, each cell having a flagellum. The origin of the Animal Kingdom involved a series of changes that caused a shift from life as a single cell to life as a miniature aquatic ball of cells.

Recall that animals are not the only multicellular organisms on Earth: plants, fungi, and slime moulds are other examples of life forms built from a multitude of cells. These groups, however, did not arise from the same ancestor. Each evolved from a different single-celled organism. The multicellular plants are not closely related to animals or choanoflagellates, and they evolved in a completely different part of the tree of life. Fungi, such as mushrooms, brewer's yeast, and athlete's foot, are nowhere near the plants, and they again evolved from their own single-celled ancestor. Perhaps surprisingly, the fungi and their ancestors fit into the same part of the tree of life as the animals and the choanoflagellates, a group called the Opisthokonta. Multicellularity evolved twice in the opisthokonts: once to give animals, and once to give fungi. It is a sobering thought that we are more closely related to mushrooms, than mushrooms are to plants.

Why should multicellularity evolve at all? After all, the vast majority of individual living things on Earth have just a single cell, including bacteria, 'Archaea', and a huge range of single-celled eukaryotes (loosely called 'protists') such as *Amoeba* and choanoflagellates. Having many cells allows an organism to grow larger, which in turn might allow it to avoid the predatory intentions of other cells, or to colonize environments not accessible to single-celled life. This may be true, but it is unlikely to be the original reason for the evolution of multicellularity. After all, the first multicellular organisms, such as the ancestor of the animals, was probably little more than a microscopic ball of cells, confined to much the same habitat and way of life as its single-celled collared flagellate relatives. The problem of the origin of animals remains unsolved, although some intriguing ideas have been put forward. One clever suggestion, proposed by Lynn Margulis, is that multicellularity allows a fundamental division of labour: some cells can divide and grow, while others carry on feeding. But why should a single cell not divide and feed at the same time? The idea is that in single-celled organisms with a flagellum, such as the collared flagellates, a key part of the cell's machinery (known as the 'microtubule-organizing centre') must be used either for moving chromosomes during cell division or for feeding by anchoring the waving flagellum, but not both at the same time.

Another imaginative model has a slightly macabre appeal. The crux of the idea, proposed by Michel Kerszberg and Lewis Wolpert, is that self-cannibalism was an initial driver for the evolution of multicellularity. Consider a population of single-celled organisms, such as collared flagellates or their relatives, together with a few mutants in which the daughter cells do not fully separate after cell division. These mutants will form clumps or colonies of cells. Both forms feed on bacteria which they filter from the surrounding water. When food is abundant, the single-celled organisms and their colony-forming sisters can all acquire nutrients and reproduce successfully. At other times, however, food may become scarce, perhaps due to a change in the environment. Inevitably, many organisms will die as they fail to obtain enough nutrients to maintain basic cellular processes. The colony-forming mutants, however, have an instant failsafe mechanism for times of hardship: cells can eat their neighbours. This could happen by sharing of nutrients, allowing just a proportion of cells in the colony to survive, or more dramatically, by some cells decaying into food for adjacent cells. The consequence is that the colony-forming mutation would have a selective advantage in times of food shortage, and more of them would survive to reproduce. Self-digestion may sound grim, but it is actually a strategy used by several animals, from flatworms to humans, in times of starvation.

The whole of the Animal Kingdom has its origins in these ancient colonies of cells. Over the past 600 million years, possibly longer, the descendants of these cell colonies diversified and radiated through evolution, giving rise to the millions of different animal species on Earth today. Animals originated in the sea, but they have since colonized fresh water, land, and air. Examples include the flatworms and fishes found in streams and rivers, snails and snakes on dry land, and butterflies and birds in the air. Some, such as flukes and tapeworms, have invaded the bodies of other animals, while a few, such as dolphins, have returned to sea again. This great diversification spans a huge size range. The parasitic myxozoans and dicyemids have shrunk and simplified so that they are no larger than tiny colonies of cells, while giant whales steer their 100-tonne bodies gracefully through the oceans. To make sense of this vast diversity, we need to focus on the most fundamental unit of classification in the Animal Kingdom: the phylum.

Chapter 2
Animal phyla

> Classifications are theories about the basis of natural order,
> not dull catalogues compiled only to avoid chaos.
>
> Stephen Jay Gould, *Wonderful Life* (1989)

Patterns and branches

For centuries, naturalists and philosophers have struggled to make
sense of the range of life on Earth. One of the earliest and most
pervasive ideas was that of a 'Scale of Nature' in which living, and
sometimes non-living, things were arranged into a linear hierarchy.
Each ascending rung on a ladder represented increasing
'advancement', based on a blend of anatomical complexity, religious
significance, and practical usefulness. The idea had its origins in
the thinking of Plato and Aristotle, but was crystallized by the
work of the 18th-century Swiss naturalist Charles Bonnet. In
Bonnet's scheme, the Scale of Nature rose from earth and metals,
to stones and salts, and stepwise through fungi, plants, sea
anemones, worms, insects, snails, reptiles, water serpents, fish,
birds, and finally mammals, with man sitting comfortably on top.
Or almost on top, being marginally trumped by angels and
archangels. It is easy to ridicule such ideas today, but Bonnet had a
good knowledge of the natural world. For example, it was Bonnet
who discovered asexual reproduction in aphids and the way that

butterflies and their caterpillars breathe. Furthermore, the idea of a Scale of Nature still pervades much modern writing, with many scientists talking of 'higher' or 'lower' animals: language that bears an uncanny resemblance to this old and discredited idea.

The dismantling of the Scale of Nature occurred gradually. A significant blow came from the respected French anatomist, palaeontologist, and advisor to Napoleon, Baron Cuvier. From his detailed studies on the internal anatomy of animals, Cuvier reached the conclusion that there were four fundamentally different ways to construct the body. These were not superficial differences, but were deeply rooted in the organization and function of the nervous system, brain, and blood vessels. In 1812, Cuvier organized the Animal Kingdom into four great branches (*embranchements*), named the Radiata (circular animals such as jellyfish, plus, surprisingly to modern biologists, starfish), Articulata (animals with a body divided into segments, such as insects and earthworms), Mollusca (animals with a shell and a brain), and Vertebrata (with bony skeletons, muscular heart, and red blood). No system was proposed to link these *embranchements*, and hence they stood parallel to each other, with equal status rather than as a hierarchy.

Cuvier, unlike his contemporary Lamarck, was not a believer in evolution. Paradoxically, however, it is evolution that provides the logical reason why Cuvier's *embranchements* could stand with equal status. As argued later by both Charles Darwin and Alfred Russel Wallace, evolution explains why every animal species has similarities with others, and why groups of species with common features can be identified. To exploit the familiar simile of evolution as a branching tree, or Wallace's more poetic phrase a 'gnarled oak', we should be able to describe small 'twigs' of closely related species embedded within larger and larger 'branches' including yet more distant relatives, all sharing common ancestry in evolution. We can then offer meaningful names to the small and large branches of the tree. The large branches within the Animal Kingdom are 'phyla' (the singular being 'phylum').

The tree analogy highlights the crux of the classification system for animals: names must reflect the natural relationships generated by evolution. Naming groups of animals is quite different from classifying a collection of inanimate objects, such as teapots, postage stamps, or beer mats. Inanimate objects can be grouped into multiple alternative arrangements, based on different properties such as colour, size, or country of origin, all of equal relevance. To classify living organisms in such a way would be to miss the fundamental point: a classification system based on evolution reflects natural order. It is a statement of relatedness, a hypothesis that proposes a particular evolutionary history.

The list of life

How many animal phyla are there? In other words, how many 'large' branches of the evolutionary tree of animals are there? This begs the immediate question of how large (or small) a branch must be to warrant being called a phylum. This is a controversial issue, but in practice animals within the same phylum should share particular anatomical structures or features different from other phyla. In James Valentine's words, 'phyla are morphologically-based branches of the tree of life'. Phylum names can never be used to group together animals from different branches, nor should one phylum be nested within another. These rules work very well for most of the Animal Kingdom, and for all of the very familiar types of animals, but there are still disputes and disagreements concerning how many phyla are needed to classify the less well-known species. One thing for certain is that Cuvier's four categories represent far too gross a simplification; the number of animal phyla most often quoted today is between 30 and 35.

In recent years, several 'new' phyla have been proposed. This sometimes happens when a phylum needs to be split in two because research has found that it mistakenly contains animals from distinct branches of the animal tree. One example is the splitting of the former phylum Mesozoa into two phyla: the phylum Rhombozoa, containing some tiny worm-like parasites, and the phylum Orthonectida, containing more tiny worm-like parasites, these ones living rather improbably in the urine of octopus and squid. A more controversial example concerns the phylum Platyhelminthes (flatworms, tapeworms, and flukes), out of which some species have recently been removed and placed in a new phylum, the Acoelomorpha. New phyla are also proposed if totally new species are found with unusual and apparently unique body structures, and which do not fall within a pre-existing phylum. Both criteria must be satisfied in order to erect a new phylum. Since the 1980s, this has occurred just a few times, notably with the discovery of the phylum Cycliophora (tiny animals living attached to the mouthparts of lobsters and scampi), the phylum Loricifera (miniature urn-shaped animals found clinging to sand grains), and the phylum Micrognathozoa (even smaller animals found in a fresh-water spring in Greenland).

Phyla also get lost. This is not through extinction, or at least we can say that no phylum has gone extinct since human records began. Instead, phyla cease to be valid when it is discovered that the whole group actually fits inside another phylum. The two groups must, logically, be merged into one. This happens surprisingly often, most usually when a group of animals with very strange anatomy was originally classified as a distinct phylum, only for later research to reveal that they are actually greatly modified members of another group of animals. The best-known example concerns the giant tube worms, or pogonophorans, which are famous inhabitants of deep-sea hydrothermal vents around the Galapagos Islands and the mid-Atlantic ridge. Considering that some pogonophorans grow to 2 metres in length, it is surprising that their evolutionary relationships proved hard to track down. However, DNA sequence data now indicate that pogonophorans are modified members of the phylum Annelida, which contains such well-known animals as earthworms and leeches. Another example concerns the former phylum Pentastomida, or tongue worms, containing large (up to 15-centimetre) scaly parasites that hook inside the nasal passages of birds and reptiles. Despite their horrific appearance, it is now clear from DNA and cell structure analyses that tongue worms are actually highly modified crustaceans, placed well within the phylum Arthropoda and close to fish lice.

Here, I recognize 33 animal phyla. Of these, 9 phyla contain well-known animals familiar to almost everybody. A further 4 phyla can be found with a little effort, in ponds, in gutters, or by a walk along the sea shore. The 9 very familiar animal phyla are: Porifera (sponges), Cnidaria (jellyfish, corals, and sea anemones), Arthropoda (including insects, spiders, crabs, and centipedes), Nematoda (roundworms such as the river blindness parasite of humans, and the slug-killer nematodes used by gardeners), Annelida (earthworms, ragworms, and leeches), Mollusca (including snails, oysters, and octopus), Platyhelminthes (flatworms, flukes, and tapeworms), Echinodermata (starfish and sea urchins), and Chordata (including fish, frogs, lizards, birds, and mammals such as humans). The 4 additional phyla to be found with relative ease are Bryozoa ('moss animals', easily seen as arrays of tiny brick-shaped boxes on the fronds of seaweed), Nemertea (squidgy, slow-moving 'ribbon worms' found under rocks on the sea shore), Rotifera ('wheel animals' present by the thousand in pond water), and Tardigrada (miniature 'water bears' found in moss and topping the cuteness list for most zoologists).

In attempts to understand how animals have diversified through evolution, how they function, and how they are adapted to particular environments, it is best to start at the level of the phylum. Since phyla are 'morphologically-based branches of the tree of life', it follows that knowing which phylum a species belongs to helps us when making comparisons to other members of the same, or a related, group, and to consider how the anatomy of that organism relates to function. For example, knowing that an animal is in the phylum Nematoda draws immediate attention to the thick elastic cuticle and pumping pharynx found in this phylum, which is relevant to understanding the lifestyle and properties of that animal. Conversely, ignoring classification leads to confusing comparisons between distantly related species, which may have very different body plans and different constraints on their evolution and way of life. But we must not consider the animal phyla simply as a list of 33 different categories. They each comprise a branch of the evolutionary tree. And, of course, branches are always connected to other branches. Because of this, some phyla are more closely related to each other than they are to others. This information is vital for understanding how structure, function, and evolution interconnect in the Animal Kingdom.

Table 1

Phylum	Where in tree?	Examples
Placozoa	Basal animals	
Porifera	Basal animals	Sponges
Cnidaria	Basal animals	Jellyfish, coral, sea anemones
Ctenophora	Basal animals	Comb jellies
Annelida	Lophotrochozoa	Earthworms, ragworms, leeches
Mollusca	Lophotrochozoa	Snails, oyster, squid, octopus
Nemertea	Lophotrochozoa	Ribbon worms
Brachiopoda	Lophotrochozoa	Lamp shells
Phoronida	Lophotrochozoa	Horseshoe worms
Bryozoa	Lophotrochozoa	Moss animals
Entoprocta	Lophotrochozoa	

Platyhelminthes	Lophotrochozoa	Flatworms, flukes, tapeworms
Dicyemida	Lophotrochozoa	
Rotifera	Lophotrochozoa	Wheel animals
Gastrotricha	Lophotrochozoa	
Gnathostomulida	Lophotrochozoa	
Micrognathozoa	Lophotrochozoa	
Cycliophora	Lophotrochozoa	
Arthropoda	Ecdysozoa	Insects, spiders, crabs, centipedes
Onychophora	Ecdysozoa	Velvet worms
Tardigrada	Ecdysozoa	Water bears
Nematoda	Ecdysozoa	Roundworms
Nematomorpha	Ecdysozoa	Horsehair worms
Kinorhyncha	Ecdysozoa	Mud dragons
Priapulida	Ecdysozoa	Penis worms
Loricifera	Ecdysozoa	
Echinodermata	Deuterostomia	Starfish, sea urchins, sea cucumbers
Hemichordata	Deuterostomia	Acorn worms
Chordata	Deuterostomia	Sea squirts, amphioxus, fish, humans
Chaetognatha	Lophotrochozoa/ Ecdysozoa	Arrow worms
Acoelomorpha	Uncertain	
Xenoturbellida	Uncertain	
Orthonectida	Uncertain	

Chapter 3
The evolutionary tree of animals

> The time will come, I believe, though I shall not live to see it,
> when we shall have very fairly true genealogical trees of each
> great kingdom of nature.
>
> Charles Darwin, letter to T. H. Huxley (1857)

Building a tree of life

Darwin realized that a branching tree was a good metaphor for
describing the course of evolution. In 1837, he made a small sketch
of an evolutionary tree in one of his personal notebooks, with the
tantalizing words '*I think*' written above. The concept would have
come quickly to Darwin once he realized that one species could
give rise to two or more 'daughter' species – a process known as
speciation. Evolutionary trees, or phylogenetic trees as they are
also called, are simply diagrams that depict these speciation events.
Every branching point on a phylogenetic tree, where one line forks
to give two lines, is a visual portrayal of one species becoming two.

Phylogenetic trees are easy to comprehend when they include
similar animal species. For example, if one line on a tree leads
to the Large White butterfly (*Pieris brassicae*) and another
leads to the Small White butterfly (*Pieris rapae*), the point where
these two lines meet marks the speciation event separating these
two very similar butterflies. This is the point in history when two

populations of their 'common ancestor' became separated such that they could no longer interbreed. Importantly, these two populations will not yet have acquired the distinct characters of the two species; indeed, they may look essentially identical. But very often phylogenetic trees do not just contain closely related species; they may depict the evolutionary relationships between large sets of animals, such as between insects, spiders, snails, jellyfish, and humans. These trees should be viewed in exactly the same way. If one line on the tree leads to insects and another leads to spiders, where these two lineages meet marks the long-extinct common ancestor of the two groups. That ancestor was neither insect nor spider, and when it underwent speciation, it gave rise to the almost indistinguishable ancestors of these two groups.

Although Darwin sketched the idea of a tree in his notebook, and enlarged upon it in his only illustration in *The Origin of Species*, he did not try to resolve who was actually related to whom. For Darwin, evolutionary trees were just a concept: a way of thinking about evolution. Many subsequent evolutionary biologists did attempt to put names onto branches of the tree. It is an important problem and one that should be soluble. After all, there should be one single tree of animal life, depicting the true course of animal evolution. Any drawing of a phylogenetic tree is therefore a clear and explicit hypothesis about the pathway followed in evolution. Some of the earliest evolutionary trees were drawn by the German zoologist Ernst Haeckel in the 1860s and 1870s. Many of Haeckel's trees were striking in their artistic detail, complete with knotted bark and twisted branches, and the names of particular animal groups at the end of each twig or leaf. Haeckel based his trees, and therefore his hypotheses about animal evolution, on several lines of evidence, but he especially liked to use characters from embryology. This was partly because he thought that embryos change slowly in evolution. Also, even when adults look very different, similar features can sometimes be found in their development. Some of Haeckel's conclusions are still compatible with modern ideas, such as his placement of jellyfish and sea anemones on a branch that split early from the rest of animal life. Other ideas seem surprising to us today and are certainly incorrect, such as placing echinoderms (starfish and sea urchins) as a branch close to arthropods such as insects and spiders.

Over the next 80 years, zoologists made better descriptions of the anatomy of animals and studied their embryonic development in more detail, with attention being paid to the great diversity of invertebrate phyla. But even by the middle of the 20th century, no clear consensus had been reached. There was no single agreed phylogeny of the Animal Kingdom. Each author would give a slightly different evolutionary tree, although certain relationships were always present. One scenario, outlined below, became prevalent particularly in American textbooks and has been called the 'Coelomata hypothesis'.

The Coelomata hypothesis

In this evolutionary tree, the main lines of evidence used to decide which animal phyla are most closely related were symmetry, germ layers, body cavities, segmentation, and patterns of cell division in the early embryo. Most familiar animals, including worms, snails, insects, and ourselves, have only one mirror-image plane, or axis of symmetry. This runs in the head-to-tail direction, separating the left-hand side from its mirror image on the right-hand side of the body. There are many deviations from precise symmetry, such as coiled shells in snails, lopsided claws in crabs, or placement of the human heart on the left-hand side of the body, but these are minor modifications. Fundamentally, most animals have near mirror-image left- and right-hand sides – an organization called 'bilateral' symmetry. In contrast, four animal phyla have no clear head and tail ends, and no left or right sides. These 'non-bilaterian' phyla, or basal animal phyla, have either no symmetry or radial symmetry. They comprise the Cnidaria (jellyfish, sea anemones, corals) and Porifera (sponges), plus two less well-known groups called Ctenophora (comb jellies) and Placozoa.

The second line of evidence was the number of 'germ layers'. Germ layers are layers of cells that arise early in the embryo and become more complex during development. Most animals have three germ layers, with the inner layer (endoderm) forming the wall of the gut, the outer layer (ectoderm) forming skin and nerves, and the middle layer (mesoderm) developing into muscle, blood, and other tissues. However, the non-bilaterian or basal phyla have only two germ layers (ectoderm and endoderm), at least to a first approximation. Whether there is something similar to mesoderm in these animals is controversial. Because of these two lines of evidence, symmetry and germ layers, the bilateral animals were placed in one large group, called the 'Bilateria' or 'bilaterians' (also called 'triploblasts' on account of the three germ layers), with the other phyla arising from branches that separated earlier in animal evolution.

Moving onto the bilaterians, one character given special attention in the Coelomata phylogeny was the presence or absence of fluid-filled spaces within the body. The embryos of some bilaterians, most notably annelids (such as earthworms) and molluscs (for example, slugs and snails), have large fluid-filled body cavities, lined with leak-proof epithelial cell sheets. The embryos of chordates, such as humans, also have these cavities, as do embryos of echinoderms (starfish and sea urchins). A body cavity of this type is called a coelom; these animal phyla were therefore called 'coelomates', and they were grouped near each other in the evolutionary tree. In earthworms, the coeloms persist into the adult, where they act as a liquid skeleton. In some other animals, including arthropods (such as insects and spiders), the coeloms can be very small or may disappear later in development, but these animals were still placed within the coelomate part of the tree. (In some other phylogenetic trees, only some of the coelomates were grouped together.) Another reason why arthropods were placed close to annelids is that both types of animal have bodies divided into repeating articulated units or 'segments'. The segmentation can be seen very clearly in the body of a centipede or an earthworm as a series of rings around (and inside) the body. In many trees, therefore, a particular super-branch of 'segmented coelomates' was defined, and called the Articulata.

In contrast to coelomates, there are also bilaterians in which the mesoderm remains solid, without any fluid-filled cavity. These animals were called acoelomates and include Platyhelminthes (flatworms, flukes, tapeworms) and Nemertea (ribbon worms). Intermediate between these two categories were the pseudocoelomates, such as Nematoda (roundworms), which have poorly defined body cavities without an epithelial cell layer. An assumption of the Coelomata phylogeny was that all the coelomates grouped together, and that the acoelomates split off earlier. The acoelomates were thought to be ancestors of the coelomates, and thus the most 'primitive' of the bilaterian animals. Another implication was that there had been a rise in complexity through bilaterian evolution, from acoelomate to coelomate, possibly via pseudocoelomate, neatly mirrored by the animal phyla surviving today.

A new tree of animals

Not every zoologist adhered to the above view, but it remained a popular hypothesis for several decades. The most prevalent alternative theory split the Bilateria into two main groups (Protostomia and Deuterostomia) and gave less attention to body cavities; however, it still used segmentation to group arthropods and annelids into Articulata. In 1988, however, a new line of evidence was brought to bear on the question, and this quickly gave an indication that something might be very wrong with the Coelomata hypothesis, and also with the idea of Articulata. A team of researchers at Indiana University in the USA, headed by Rudolf Raff, set out to use gene sequence data to investigate the evolutionary relationships between animal phyla. Genes accumulate mutations over time, so differences in DNA sequence between species should reflect how long it has been since they shared an ancestor. Animal phyla that are close relatives would have similar DNA sequences for a particular gene; more distant relatives would have more different DNA sequences. Raff and colleagues focused attention on the gene coding for small subunit ribosomal RNA, one of the components of the ribosome, a structure found in all cells. The main advantage of this gene is that it is present in every animal species, doing the same job: helping make proteins.

The 1988 study marked the beginning of a revolution in using DNA sequence information to hunt for the true phylogenetic tree of animals. Even though the technology was new and the analysis methods in their infancy, one conclusion seemed clear right from the start. The segmented annelids and segmented arthropods had very different ribosomal RNA gene sequences; there was no evidence supporting the Articulata group. Over the next 20 years, the DNA sequences of many more genes were determined, from many more species, and computer-based analysis methods were refined and improved. The most reliable phylogenetic trees now include over a hundred genes from each animal, and they show a remarkably consistent picture. This 'new animal phylogeny' has some similarities with the older trees, but it also has some key differences.

In the new animal phylogeny, the four non-bilaterian phyla branch off the main tree early, just as they did in the Coelomata and other morphology-based phylogenies. This implies that germ layers and symmetry were giving an accurate picture. Jellyfish, sea anemones, corals, comb jellies, and sponges are indeed 'basal' animals. After these basal phyla diverge, the rest of the animals comprise the Bilateria. It is within this bilaterian group where the hypotheses differ. For example, in the new animal phylogeny there is no region of the tree composed only of acoelomates, no section for pseudocoelomates, and no grouping of just coelomates. Instead, all three types of body organization are mixed up.

Since coeloms are found in several different parts of the 'new' tree, this implies either that body cavities arose more than once in evolution, or that they can be lost, or both. From a functional point of view, this is perhaps not surprising. Fluid-filled cavities provide an advantage to invertebrates living in many environments – they provide support to the body and act as an incompressible bag against which different sets of muscles can squeeze. For soft-bodied animals, this increases the power and efficiency of animal movement, allowing burrowing, faster crawling, and even occasionally swimming. From the perspective of drawing evolutionary trees, it means that body cavities were poor markers of relatedness. The same is true for segmentation. Dividing the body into units offers advantages in certain environments, for example by increasing the efficiency of movement, and it also probably evolved more than once. Segments, like coeloms, come and go too readily in evolution to be markers of who is related to whom. It seems there is no grouping called the Coelomata, and no Articulata.

So what shape is the evolutionary tree built from DNA sequences? In the new animal phylogeny, rapidly gaining wide acceptance, the Bilateria divide into three great groups, which might be called 'superphyla'. Each of these contains several phyla. The superphylum to which we belong is called the Deuterostomia. Along with our phylum the Chordata, this deuterostome group includes the Echinodermata (starfish and sea urchins) and the Hemichordata (including the foul-smelling acorn worms). The older phylogenies almost always had a group called Deuterostomia as well, but it usually also included a few other animals which have now been moved elsewhere on the basis of the DNA data, notably the Chaetognatha, or arrow worms.

The other two great superphyla of bilaterians were a surprise. They had not been suspected from comparison of anatomy, and were not present in any of the older, traditional trees. Nonetheless, each is now strongly supported by DNA sequence data. Since they were proposed only recently, these two groups of animals needed new names. They are each a bit of a mouthful. One, containing Arthropoda (insects, spiders, crabs, centipedes), Nematoda (roundworms), and several other phyla, is called 'Ecdysozoa'. The other, containing Annelida (earthworms, leeches), Mollusca (snails, octopus), Platyhelminthes (flatworms, flukes, tapeworms), Bryozoa (moss animals), and others, delights in the name 'Lophotrochozoa'.

A phylogenetic tree is best shown using a diagram. As summarized in Figure 2, the new animal phylogeny has the four non-bilaterian

The Animal Kingdom

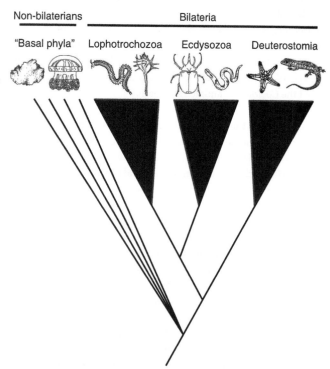

2. The 'new animal phylogeny' based on DNA sequence data

phyla branching early in animal evolution, leaving the large group of Bilateria. The bilaterians then divide into the three great superphyla, Deuterostomia, Ecdysozoa, and Lophotrochozoa, as described. Incidentally, the latter two groups are closest to each other, and approximate to the 'Protostomia' of some older trees. It is important to realize that among the three great groups, none is 'higher' or 'lower' than any other, since all still exist today. There is no rising Scale of Nature. In the remaining chapters of the book, we will look at the animals in each of these branches, starting with the non-bilaterian phyla, and then dealing with the three large bilaterian superphyla in turn. The order is arbitrary. Just because humans sit within the Deuterostomia does not give our group any special priority in the tree.

Chapter 4
Basal animals: sponges, corals, and jellyfish

The bottom was absolutely hidden by a continuous series of
corals, sponges, actiniae and other marine productions, of
magnificent dimensions, varied forms and brilliant colours....
It was a sight to gaze at for hours, and no description can do
justice to its surpassing beauty and interest.

Alfred Russel Wallace, *The Malay Archipelago* (1869)

Porifera: the sponges

Sponges are the least animal-like of all members of the Animal
Kingdom. Most sponges are vase-shaped, but some appear as
lumpy, irregular growths encrusting rocks in the sea, or pebbles and
fallen branches in lakes and rivers. For these animals, the concepts
of front and back, top and bottom, or left and right, do not apply
rigidly. They have no clear nerve cells or muscles, but they can move,
very slowly, and – like other animals – sponges can respond to touch
and can sense chemical changes in their environment. Unlike other
animals, they do not have a true mouth or gut, but instead use a
complicated system of water flow to capture food. Sponges can be
recognized by the presence of one or more large pores or holes on
their surface, although there are also thousands of much smaller
pores. A constant stream of water flows into the small holes and out
of the larger ones. This water current, which carries dissolved

oxygen and particles of food such as bacteria, is set up by an important type of cell found lining a network of hollow canals and cavities within the sponge. These feeding cells, or choanocytes, have a beating flagellum and resemble the unicellular collared flagellates encountered earlier. They function in a different way, however, since – unlike collared flagellates – the choanocytes do not catch food using their collar as a simple net. Instead, the chambers containing choanocytes have a larger cross-sectional area than the pores, meaning that the water flow slows greatly once it is drawn into the sponge. With the incoming water flow now almost stationary, sponge cells can engulf bacteria and other particles of food.

Although sponges have many different types of cells, most are not organized into organs with discrete functions, like kidneys, livers, or ovaries (although the choanocyte chambers could be considered simple organs). For this reason, sponges are sometimes described as having a 'tissue-level' organization. Some sponges have astonishing powers of regeneration, so extreme that they were the inspiration for regenerating aliens in the science-fiction television series *Doctor Who*. The defining experiments that revealed this property were published in 1907 by Henry Van Peters Wilson of the University of North Carolina, USA. Wilson mashed up a living sponge and passed it through a fine cloth, the sort used for sieving flour, thereby splitting most of it into individual cells. Wilson then observed that these cells gradually crawled back together and reassembled into a new sponge! Furthermore, if the cells of two different species were mixed together, they would sort themselves out and regenerate into the two original sponges again. Although regeneration is found in many branches of the Animal Kingdom, no other animals are as expert as some of the sponges.

In between the outer and inner layers of cells, sponges have 'connective tissue' reinforced with tough fibres of a protein called spongin or with minute spears or stars (spicules) made of calcium carbonate or silica. The former type of sponge, with a spongin skeleton but no spicules, was the source of the old-fashioned bath sponges widely used for washing and cleaning, although now largely replaced by synthetic foams. Examples include the genera *Spongia* and *Hippospongia*. The collection and use of sponges dates back many centuries. In the 1st century AD, Pliny the Elder described in detail how to use sponges to clean wounds, reduce swellings, arrest bleeding, and treat stings. Even earlier, in the 4th century BC, Aristotle described which type of sponge should be used to line helmets, writing:

> the sponge of Achilles is exceptionally fine and close-textured and strong. This sponge is used as a lining to helmets and greaves, for the purpose of deadening the sound of the blow.

Remarkably, it is not only humans who use sponges as tools. In Shark Bay on the Western Coast of Australia, a population of bottlenose dolphins has learned to snap off pieces of living sponge and attach these to their snouts to protect themselves when foraging for food in the sandy bottom.

The sponges comprise a phylum, the Porifera, which is in turn divided into three classes: Demospongia (including bath sponges), Calcarea (with calcium carbonate spicules), and the rare, deep-sea-dwelling Hexactinellida. The hexactinellids, also known as glass sponges, are particularly beautiful and have some important differences from the other sponges. One peculiar feature is that much of their body is made of 'syncytia': sheets of cytoplasm containing many nuclei, not separated into individual cells by membranes. Also unusual is that their silica spicules are woven together into delicate lattice-like structures, like elaborate cages of glass. The best-known example is 'Venus's flower basket', *Euplectella aspergillum*, which lives attached to rocks on the Pacific Ocean floor, and has a cylindrical 30-centimetre tower-like skeleton made of intricately laced glass fibres. Often, it is found with a pair of live shrimps, male and female, trapped within the glass fibre cage, having grown too large to swim out through the gaps in the sponge's skeleton. The shrimps' offspring, however, can escape through the latticework walls and swim off to populate other Venus's flower baskets, leaving the two parent shrimps behind in their permanent partnership. In an old Japanese custom, specimens of this sponge were often given as wedding gifts symbolizing eternal union.

The strange case of *Trichoplax*

Sponges are not the only animals to lack the three defined axes: head to tail, top to bottom, and left to right (bilateral). Three other phyla are also 'non-bilaterian' in their organization: the Cnidaria (sea anemones, corals, and jellyfish), the Ctenophora (comb jellies), and the Placozoa. Originally, only a single species was placed in the last of these phyla, a tiny pancake-shaped creature called *Trichoplax adhaerens*, meaning 'sticky, hairy plate'. Recent genetic analysis suggests it is not alone, however, and there are actually several similar species of these minute creatures, crawling and floating in tropical and subtropical seas from the Pacific Ocean to the Caribbean, and the Mediterranean to the Red Sea. At first sight, *Trichoplax* could easily be confused for a very large amoeba, between half and one millimetre across, but on closer examination it can be seen to be made of thousands of cells, as befits a true animal. Its irregular flattened shape has no preferred front end, and it crawls along hard surfaces in any direction through a combination of shape changes and the beating of thousands of microscopic cilia covering its underside. With no mouth or gut, *Trichoplax* feeds by secreting enzymes from its undersurface, which break down food matter, such as single-celled algae, into nutrients for absorption. All in all, placozoans are very unusual animals and they have long puzzled zoologists.

They were first discovered in 1883 by the German zoologist and sponge expert Franz Eilhard Schulze but, interestingly, he did not actually discover *Trichoplax* in nature. Schulze found the new animal crawling on the glass walls of a marine aquarium in Austria, meaning that at first there was no clue as to where it lived

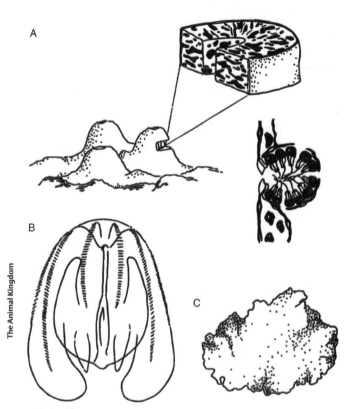

3. A, Porifera, or sponge: *Haliclona*, showing structure of choanocyte chamber; B, Ctenophora, or comb jelly: *Mnemiopsis*; C, Placozoa: *Trichoplax*

in the wild. Indeed, many zoologists later claimed that Schulze was mistaken in describing *Trichoplax* as a new animal at all, arguing that it was simply the larva of a well-known sea anemone-like animal. It was almost a century before Schulze was fully vindicated, as extensive research on placozoans in the wild and in the laboratory have now proved them to comprise a distinct phylum in their own right, albeit one rather short on species numbers.

Those wishing to follow Schulze should be warned, however. I was once forcibly told to leave an aquarium shop when the irate manager found me peering through a magnifying glass at the scum in his fish tanks.

Ctenophora: comb jellies

The Ctenophora comprise a third phylum of non-bilaterian animals, and one very different in body organization from either sponges or placozoans. Also known as comb jellies, the ctenophores are slow-moving predators that drift through the seas eating other slow-moving animals, such as other ctenophores, crustaceans, and marine larvae. Unlike most predators, comb jellies do not chase or stalk their prey. They simply bump into small planktonic animals and ensnare them with tiny drops of glue secreted by specialized cells, usually concentrated along two long tentacles trailing away from the sides of the mouth. Ctenophores, in contrast to sponges and placozoans, have nerve cells and a balance sense organ so they can interact with their environment rapidly and responsively. Although most are essentially blobs of jelly just a few centimetres in size, almost everyone who has seen one alive would rank comb jellies amongst the most beautiful animals on the planet. Their most conspicuous features are the eight 'combs' that run as strips along the body, each containing thousands of cilia. The cilia beat in a highly coordinated way, with each one beating just after its neighbour to form a set of 'metachronal waves', rather like the 'Mexican waves' that occasionally circulate around a football stadium during lulls in play. This gentle wafting of thousands of tiny cilia propels the animal slowly and silently through the sea, but also scatters light to create a shimmering rainbow of colours, constantly changing and flickering. The best-known comb jellies are the grape-sized 'sea gooseberries' such as *Pleurobrachia* found throughout Pacific and Atlantic oceans, and around the British coast. But the most spectacular comb jelly is undoubtedly the giant 1-metre-long *Cestum veneris*, or Venus's girdle, named after the Roman goddess of love. Instead of the usual egg-shape typical of ctenophores, this striking, iridescent animal has an elongated, ribbon-like body that shimmers in the sea as the sun's rays are scattered by its rows of cilia. In Richard Dawkins's words, *Cestum* is 'too good for a goddess'.

Most comb jellies have little direct effect on humans, except through a minor role in marine food webs. One species, however, stands apart as the villain of the basal invertebrates. In the 1980s, the Atlantic comb jelly *Mnemiopsis* was accidentally introduced into the Black Sea, probably in ballast water carried by commercial shipping. Once in its new environment, away from natural competitors and predators, it reproduced rapidly – all the while consuming vast quantities of larval fish and crustaceans. Some (controversial) estimates placed the total seething mass of diminutive comb jellies in the Black Sea at over half a billion tonnes. The local anchovy fishery, already under heavy fishing pressure, was decimated. While ecologists debated what to do, a possible solution arrived, unplanned, in the shape of another accidental introduction. The newcomer was a second comb jelly, this time the voracious *Beroe*. Fortuitously, *Beroe* does not eat fish or crustaceans, but is instead a specialist predator of other comb jellies and nothing else. As the invading *Beroe* now feast on the *Mnemiopsis*, the fish stocks are showing gradual and welcome signs of recovery.

Cnidaria: stings and super-organisms

Of the four 'non-bilaterian' phyla, sponges and placozoans lack precise symmetry, while comb jellies have 'biradial' symmetry, meaning that their bodies are symmetrical by a 180-degree rotation. The fourth and largest of the non-bilaterian phyla, the Cnidaria, contains some very familiar animals including jellyfish, sea anemones, and corals. The bodies of these animals also lack head-to-tail, top-to-bottom, and left-to-right axes, and with few exceptions they also have radial or rotational symmetry. The basic structure of a cnidarian body is cup-shaped or flask-shaped, with a single large opening at one end that acts as both mouth and anus. Tentacles surround this opening, each armed with thousands of stinging cells called cnidocytes. These cells, which fire tiny barbed harpoons or nematocysts laced with poison within three milliseconds of being touched, are the cnidarian's chief weapon of attack and defence.

Cnidarians have nerve cells, and as in ctenophores, these are arranged in a lattice-like network around the body rather than being organized into a single distinct brain and 'central nerve cord' as in most other animals. The cell layers that form the body, the ectoderm on the outside and endoderm on the inside, are separated by a substance called mesoglea. Although much of the mesoglea is made of proteins rather than layers of living cells, in many cnidarians it has scattered cells crawling within it, and in some species it even has muscle cells organized into contractile fibres. However, the cells within the mesoglea do not form complex organs, and so cnidarians are usually described as having a body built from just two basic cell layers.

The cnidarians divide into four groups. The first, the anthozoans, includes sea anemones, such as the brightly coloured Snakelocks and Beadlet anemones found in rock pools. In these animals, the single opening to the body faces upwards, while the opposite end sticks rather weakly to the rocks. Once the incoming tide has covered them, sea anemones open their crown of tentacles and wait for small prey animals to drift or swim near them, whereupon these will be promptly stung and eaten. Although generally static animals, sea anemones are not totally fixed but can detach and move to a new location by drifting or gentle swimming. They can also creep slowly along on their single adhesive foot, sometimes to find a more favourable location for feeding and sometimes to engage in ferocious slow-motion battles when two sea anemones attempt to sting each other with inflated clubs armed with nematocysts.

Corals are also anthozoans and they demonstrate a character that has arisen repeatedly in animal evolution: coloniality. Coral consists of thousands or even millions of tiny animals, each like a miniature sea anemone just a few millimetres across, but interconnected to make a giant 'super-organism'. A living coral grows by budding of the tiny 'zooids' such that the whole colony has the same genetic make-up. It is one large clone. In some species, the colony resembles a fan; in others, it branches like deer's antlers; and yet others grow to look like grisly body parts such as brains or 'dead-men's fingers'. The most impressive of all, however, are the reef-building corals which secrete calcium carbonate around the budding zooids to form giant chalky structures upon which many other animal species can also make their home.

The second group of cnidarians, and looking superficially like sea anemones, are the hydrozoans. These include some large and colourful marine species, plus the diminutive hydra found in ponds and rivers. Named after a multi-headed water beast of Greek mythology, the hydra body is a miniature tube a few millimetres in length, with an upward-facing mouth surrounded by stinging tentacles. All species of hydra will catch and eat tiny fresh-water invertebrates, but several species back this up with an additional trick. The 'green hydra' has enslaved a unicellular alga, which grows inside the hydra's gut cells, giving the whole body a bright green appearance and providing the hydra with food through photosynthesis. Just as some anthozoans live in interconnected colonies, so too with some hydrozoans. The infamous Portuguese Man o'War *Physalia physalis* is a giant colonial hydrozoan comprising a gas-filled float beneath which are thousands of connected zooids dangling menacingly in 10-metre strings, bristling with venomous nematocysts.

When a cnidarian has a 'mouth up' orientation, as in adult sea anemones, corals, and hydra, it is called a 'polyp'. The opposite body orientation, with the opening facing downwards, is known as a 'medusa' and is the form typically seen in scyphozoans or jellyfish. Many cnidarians have life cycles that alternate between the two orientations, mouth up and mouth down. There are additional differences beyond simply orientation, and recent research using gene expression patterns has shown that downward-pointing medusa tentacles are not actually the same structures as upward-pointing polyp tentacles. Jellyfish, like all cnidarians, are predators. These bell-shaped gelatinous animals drift or gently swim through the upper waters of the sea, propelled by rhythmic pulsations of their body wall. Surface waters of the oceans teem with planktonic life, such as crustaceans and immature fish, which the jellyfish ensnare with trailing tentacles armed with poisonous nematocysts. Many swimmers have had first-hand experience of accidentally brushing against jellyfish tentacles and suffering a painful rash. Several variations on the basic jellyfish theme have evolved, one of the most unusual being seen in animals of the order Rhizostomae. In these jellyfish, there is no single downward-pointing mouth, since this is closed off by fused tissue, and instead there are a multitude of tiny mouth-like openings on eight branching arms, each connected to the gut by a complicated canal system. Many rhizostomids, such as *Mastigias papua*, supplement their food intake by harbouring in their tissues millions of symbiotic algae capable of producing energy by photosynthesis. This enables *Mastigias* to live at incredibly high densities. In 'Jellyfish Lake' on the Pacific island of Eil Malk in Palau, intense aggregations of *Mastigias papua* can sometimes reach a thousand 6-centimetre animals per cubic metre of sea water.

More dangerous to humans than true jellyfish, or even the Portuguese Man o'War, are the animals comprising the fourth group of cnidarians: the cubozoans. Also known as box jellies on account of their shape, these animals are commonest in coastal tropical seas. Unlike true jellyfish, each cubozoan has 24 eyes, including 6 with lens, iris, and retina capable of forming an image of distant objects. Some species, such as the Australian sea wasp

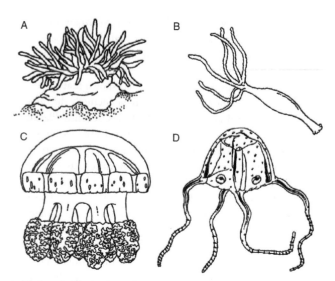

4. Phylum Cnidaria: A, Anthozoan, sea anemone; B, Hydrozoan, *Hydra*; C, Scyphozoan, or jellyfish, *Mastigias papua*; D, Cubozoan, or box jelly, *Carukia barnesi*

Chironex fleckeri, are deservedly feared by swimmers on account of their extremely powerful venom. Sea wasp stings are intensely toxic, and can be fatal even for humans. The stings from some other cubozoan species are less painful on contact, yet can trigger an unusual delayed reaction known as 'Irukandji syndrome', named after the Australian Aborigines from the north Queensland coast where box jellies are common. Swimmers stung by Irukandji cubozoans gradually develop excruciating back pain, muscle cramps, nausea, increased blood pressure, and a range of disturbing psychological effects including 'a feeling of impending doom'.

Chapter 5
The bilaterians: building a body

Man is but a worm.

Edward Linley Sambourne, *Punch Magazine* (1881)

Life with a front end

You are a bilaterian. So are fish, birds, worms, squid, cockroaches, and millions of other animals. In fact, most animals are Bilateria. As the name indicates, this vast division of the Animal Kingdom comprises the animal phyla with 'bilateral symmetry', which means that these animals have just a single line of mirror-image symmetry running right down the centre of the body. This symmetry line separates the left- and right-hand sides of the body, and by implication there must then be distinct front and back ends, and top and bottom surfaces, which are not symmetrical. In humans, the left- and right-hand sides are just as you recognize them, but the front end (anterior) of your body is actually your head, the back end (posterior) of your body is what you sit on, your top or 'dorsal' surface is found along your spine, while your bottom or 'ventral' surface is your belly. These orientations make sense when we remember that humans stood upright only recently, in evolutionary terms.

Bilateral symmetry contrasts with the rotational symmetry found in most cnidarians and comb jellies, and the lack of clear symmetry seen in placozoans or sponges. The similarity in body organization amongst the bilaterians is more than skin deep. The bilaterian animals have well-defined blocks of muscle which can be used for active movement, and almost all have centralized nerve cords with an anterior brain, plus specialized sense organs concentrated at the front end. Most have a tube-like or 'through-gut' with a separate mouth and anus, allowing efficient processing of food, and the exceptions, with just a single opening to the gut, could have reverted to this condition secondarily. The evolution of bilaterian animals marked the rise of animals with active, powerful, and directed locomotion, able to burrow, crawl, or swim while facing their environment head-on with batteries of sense organs, leaving their waste products behind them. The bilaterians truly explore and exploit the world in three dimensions.

The distinctions between bilaterians, or triploblasts as they are also known, and the more 'basal' animal phyla have been noted for over a century. In 1877, the influential English zoologist Ray Lankester contrasted the embryos of bilaterians with those of cnidarians and sponges, and pointed out that in their early development bilaterians have an extra layer of cells destined to develop into the well-defined muscle blocks of the adult. The similarities seen within embryos and in body symmetry are certainly fundamental. Towards the end of the 20th century, biologists were astounded to find similarities that went far, far deeper – right down to the DNA. The discovery that all bilaterians use the same set of genes to build their bodies represents one of the most fascinating scientific breakthroughs of the 20th century, and one that has changed biological science from the 1980s onwards. It was a discovery with explosive impact, yet this was a revolution with a slow fuse.

Homeosis and Hox genes

William Bateson is remembered today as one of the founders of the science of genetics. Long before he became famous, and soon after completing his undergraduate degree, the young Bateson published a series of scientific papers on the anatomy of the acorn worm, a marine invertebrate whose evolutionary position was then unresolved. Bateson's work was received with some acclaim, yet he was not satisfied, saying that it gave little insight into how evolution actually worked. In a letter to his mother, Bateson wrote:

> Five years hence no-one will think anything of that work, which will be very properly despised. It hasn't any bearing whatever on the things we want to know. It came to me at a lucky moment and was sold at the top of the market.

What Bateson really wanted to know was how variation arose within species. So, for the next eight years, Bateson devoted himself to cataloguing 'variants' in animals and plants, publishing in 1894 his magnum opus *Materials for the Study of Variation*. Amongst the many gems in this book, Bateson discussed an unusual type of variation in which an animal is found with one 'structure' replaced by another that would usually be found elsewhere in the body, such as an antenna growing where an eye should be. These strange 'homeotic' variants remained little more than curiosities until 1915 when Calvin Bridges showed that one such change in a fruitfly was passed on to the fly's offspring. The inheritance was key. It pointed the finger at genes. The implication was that there must be genes instructing body parts to develop correctly, and when one of these genes has an error – a mutation – the instructions are misread. One region of the body will develop as if it is a different region. In this first homeotic mutation, wings, or rather parts of wings, grew where they should not be. Bridges called the mutation *bithorax*. Oddities such as *bithorax* are far too dramatic to play any direct part in evolution; flies with such big changes to anatomy would not survive in nature. But the mutation gives a clue to how genes build bodies, and that certainly is relevant to understanding how animal evolution works.

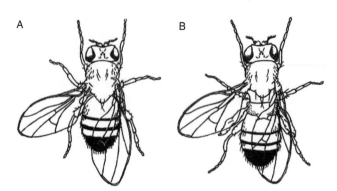

5. A, Normal fruitfly; B, Calvin Bridges's *bithorax* mutation

The finding was taken up and pursued with great energy and persistence by another geneticist, Ed Lewis. In a series of brilliant papers, including a Nobel Prize-winning masterpiece in 1978, Lewis showed that *bithorax* was not alone. There were several genes that could mutate to cause homeotic mutations, with each mutation affecting a different region of the fly's body, and all mapping to one part of one fruitfly chromosome. Another geneticist, Thom Kaufmann, found genes that controlled development of the head and front of the body, and so a picture emerged of a whole set of 'homeotic genes', each telling cells in the embryo where they were. The homeotic genes act like postcodes, telling cells where they are along the head-to-tail axis of the fly.

When their DNA was analysed, all homeotic genes turned out to be similar to each other, particularly along a stretch of 180 base pairs. This region, which became known as the 'homeobox', was a molecular signature of homeotic genes – or Hox genes as they were soon named. Homeobox sequences were also found in the DNA of some other fruitfly genes, but always in genes that had something to do with controlling development, such as the gene *fushi tarazu* involved in formation of the fly's segments. But very soon, this became a story with relevance far beyond fruitflies. Few biologists were prepared for the impact of what was unravelling. The year was 1984, the centre of activity was Basel, Switzerland, and a dynamic team of researchers including Bill McGinnis, Mike Levine, Atsushi Kuroiwa, Ernst Hafen, Rick Garber, Eddy De Robertis, Andres Carrasco, and Walter Gehring were pushing back the boundaries of biological knowledge. McGinnis and colleagues tested whether homeobox sequences could be detected in DNA extracted from other animals, with striking results. Not only did other insects have homeoboxes, but the first experiments suggested that perhaps so did worms, snails, and possibly even mice and humans! Carrasco, McGinnis, Gehring, and De Robertis quickly isolated and sequenced the DNA for a frog homeobox gene and proved the point: vertebrates really do have homeobox genes.

The buzz around the scientific community was electrifying. The top journals clamoured to showcase each new finding, and each published paper was devoured by an eager readership. Every conference and seminar was dominated by homeoboxes. After one scientific lecture in London at which a colleague reported a new developmental control gene, I recall the first question asked was 'Does it have a... shall I say the magic word?' I even knew several scientists who just stopped their life's research there and then, and started afresh to work on homeobox genes.

The discovery of the homeobox, and the finding that homeobox genes are present in animals as diverse as flies and frogs, started a revolution. Prior to 1984, there was essentially no knowledge of how genes controlled body patterning in most species. Perhaps homeobox genes provided a new way into the problem? This question prompted Jonathan Slack to liken the discovery of the homeobox to the finding of the ancient Rosetta Stone, unearthed in Egypt in 1799, which provided the first translation between ancient scripts. In the same way, did we now have a way of comparing the control of embryonic development between widely different species? The optimism was not shared by everyone, but it turned out to be well founded. Many of the homeobox genes in vertebrates, such as frogs and humans, are indeed equivalent to the fruitfly's homeotic or Hox genes, and they play essentially the same roles. As in flies, our own Hox genes act as postcodes, telling human cells where they are along the head-to-tail axis.

For evolutionary biology, the implications were immense. If vertebrates and insects have Hox genes, then surely by implication so must all Bilateria? If vertebrates and insects use these genes to denote position along the main axis, then this property too must date back to the origin of the bilaterally symmetrical animals. We can be confident in these statements simply because the common ancestor to flies and humans was also the common ancestor to all of the Ecdysozoa, Lophotrochozoa, and Deuterostomia. One bilaterian phylum might possibly have branched off a little earlier, the Acoelomorpha, but recent evidence suggests that even in these animals the Hox genes play a similar role. So this entire swathe of the Animal Kingdom, the 29 animal phyla with a clearly distinct front end and back end, the active explorers of the three-dimensional world, use the same set of genes to pattern the main head-to-tail axis.

Up and down, left and right

What of the other two body axes: top to bottom and left to right? Here too, genes have been found that ensure cells know where they are. Furthermore, just as was shown for Hox genes, it turns out that very different bilaterian animals use essentially the same genes – but with an interesting twist. In the embryo of a fly, cells on the bottom or ventral side will form the main nerve cord, with the gene *sog* playing a key role. Cells on the top or dorsal side form epidermis, with this opposite fate controlled by the gene *dpp*. Vertebrates also have *sog* and *dpp* genes, although known by different names. The vertebrate *sog* gene, called *chordin*, is expressed on the side destined to become dorsal, where our nerve cord lies; *BMP4*, one of the vertebrate *dpp* genes, marks ventral. In terms of orientation, it is simply the opposite way round to flies.

Comparing these genes more widely, it turns out that most animals are oriented like flies; it is our phylum, the Chordata, which is upside-down. Less is known about the evolution of the left–right axis, but we do at least know that two genes, *nodal* and *Pitx*, are involved in patterning this axis in animals as different as snails and humans.

The similarities are not restricted to the orientation of the body, but go right inside. For example, several of the genes controlling heart formation in vertebrates are also found in insects, where they too control development of a pulsating muscular tube. There are networks of genes specifying where an eye will form, and most of these genes are also the same whether the animal is a fly, a worm, or a human. Putting these remarkable findings together, it seems that the long-extinct ancestor of all bilaterian animals had a system of genes for telling dorsal from ventral, for distinguishing left from right, for telling cells where they are along the head-to-tail axis, and for building various internal structures and sense organs. These genes and their roles have been retained for hundreds of millions of years, albeit with some modifications, and with one of our ancestors turning upside-down for some reason. The French naturalist Étienne Geoffroy St Hilaire argued as much in 1830, but based on rather imaginative anatomical comparisons and a dubious idealistic basis. Geoffroy said 'There is, philosophically speaking, only a single animal.' His views were never accepted in his lifetime but, at least with regard to the bilaterians, it seems he may have been right.

The ancient set of genes used for building the body is sometimes referred to as the 'developmental toolkit'. Some of the toolkit genes, such as *Pitx* and the Hox genes, code for proteins that bind to DNA, switching batteries of other genes on or off. Others code for secreted proteins that transmit signals between cells, for example *nodal* and *dpp*, or interfere with signals, for example *sog*. These examples are just the tip of the iceberg, however, as the toolkit includes hundreds of genes coding for DNA-binding proteins, dozens producing secreted factors, and others coding for receptors to which the secreted factors bind. All can be found across diverse bilaterian phyla, though sometimes individual genes have been lost in the evolution of particular animal groups. The roles of the genes are often similar between different phyla, as in the examples above, but in other cases, toolkit genes have been recruited for different roles in divergent taxa. These are the ancient genes used to build the bilaterian body. But when did they arise? Does the evolution of the developmental toolkit tell us anything about the earliest steps in animal evolution?

Delving into the genome sequences of the non-bilaterian groups – sponges, placozoans, cnidarians, and comb jellies – reveals a rich picture. Some key genes are found in all animals, but several are not. Cnidarians, which might be the closest of the non-bilaterians to the bilaterians, have most of the toolkit genes, although their Hox cluster is less complex. The other phyla lack more of the toolkit genes; sponges lack Hox genes completely, for example. Stepping outside the animals, to the choanoflagellates, we see a bigger difference, with many of the toolkit genes absent. The conclusion is clear. The basic set of genes necessary to build animal bodies evolved around the time that multicellularity originated, but this set of genes was then expanded and elaborated through the earliest stages of animal evolution. A large toolkit of developmental genes was in place by the dawn of the Bilateria, half a billion years ago. Today, these genes are used to shape and pattern the myriad of animal bodies across all three great groups of bilaterians: the Lophotrochozoa, the Ecdysozoa, and the Deuterostomia.

Chapter 6

Lophotrochozoa: wondrous worms

> It may be doubted whether there are many other animals which have played so important a part in the history of the world as have these lowly organized creatures.
>
> Charles Darwin, *The Formation of Vegetable Mould through the Action of Worms* (1881)

Annelida: living ploughs and bloodsuckers

Just a year before his death, Charles Darwin published his last book. The work was enthusiastically received and, at least initially, sold even faster than had *The Origin of Species*. The unlikely bestseller, *The Formation of Vegetable Mould through the Action of Worms with Observations on their Habits*, included insights drawn from Darwin's own practical research, conducted on and off over 40 years. Now a grandfather and feeling his age, he was keen to publish his findings about earthworms 'before joining them', as he put it. The book's most important conclusion was that earthworms should not be despised as pests that left unsightly casts on well-manicured Victorian lawns, but rather they were 'living ploughs' crucial to the health of soil. Darwin showed that earthworms drag organic matter such as leaves underground, that their tunnels aerate the soil and provide channels for water drainage, and that their actions mix the soil preventing layers from becoming compacted, thereby promoting plant growth.

Earthworms even have effects on geology, through attrition of rocks and stones, and on archaeology by burying ancient remains.

Earthworms belong to the phylum Annelida and their anatomy is central to why these animals have such an impact. Annelids are soft-bodied, muscular, and elongate, with a mouth at one end and an anus at the other. They have a series of fluid-filled spaces, or coeloms, along the body, providing a degree of rigidity through internal water pressure, coupled with extreme flexibility. All these features are helpful for squeezing through spaces in soil, tunnelling, or even shifting soil by passage through the gut. But the one feature of annelids that is most crucial to the process is undoubtedly division of the body into a series of units or rings. It is a character obvious at a first glance and one that gives annelids their common name of 'segmented worms'. Splitting the body into segments, each with its own muscles, coeloms, and nervous control, allows earthworms to contract some parts of the body, making them long and thin, while at the same time other parts are squeezed lengthways to become short and fat. Thin parts probe forward into crevices, while fat parts anchor the worm still, and by passing waves of contraction from head to tail, the body is pushed forward through the soil.

There are more than 15,000 species of annelid worm, with most of these living in the sea and fresh water, rather than on land. Throughout their evolutionary diversification, segmentation has been key. For example, the predatory marine ragworms contract some segments on their left side and others on their right, contorting the body into moving side-to-side waves. This rapid but coordinated wriggling propels the animal forward at speed, enabling it to chase and catch its prey. Other marine annelids are more passive, living sedentary lives in burrows or tubes, filtering particles from the sea water. But even these worms deploy segmentation to great effect, as coordinated waves of contraction are used to flush out their burrow or tube and bring in fresh, oxygen-rich water.

One well-known group of annelids, however, has all but lost segmentation, and for good reason. This group is the Hirudinea, better known as leeches. Some leeches are predators, devouring small aquatic invertebrates. Others, as every tropical explorer or movie fan knows, feed by latching onto the flesh of larger animals and sucking their blood. These parasitic leeches attach to skin using a strong sucker, inject a powerful anticoagulant to stop blood clotting, and slice at flesh using three blade-like jaws. Since food sources such as horses, deer, fish, and human legs come along quite infrequently, leeches are adapted to take in giant meals when they get a chance, and one such modification affects segmentation. Leeches evolved from aquatic annelids not dissimilar to earthworms, but have lost the walls, or 'septa', that separate the segments internally. This allows the body to stretch as the leech gorges on blood, distending its body like a balloon. The downside of this modification is that leeches cannot manage the same coordinated waves of contraction used by earthworms or ragworms, and most species resort to less efficient looping movements.

The capacity of some leeches to drink human blood, rather painlessly, has been exploited by medicine for centuries. Two thousand years ago, the Greek physician Themison of Laodicea wrote of using leeches for bloodletting, and the practice continued in many parts of the world until well into the 19th century. Leeches provided a way to draw off 'bad blood' and correct 'imbalanced humours', seen as the source of many ailments for which the true cause was unknown. Even the word 'leech' derives from 'loece', the Anglo-Saxon word for 'physician' or 'to heal'. The demand for leeches in the 18th and 19th centuries was huge, with natural populations of the large medicinal leech, *Hirudo medicinalis*, being so over-exploited that the species is still rare today across much of Europe. Leech farms sprang up, but even these could not keep pace with demand. This is hardly surprising, since by the 1830s, an astonishing 40 million leeches were being imported into France each year. But leeches are not confined to the history of medicine, and their use has made a surprising comeback in recent years. A common complication during modern microsurgery is venous insufficiency, which occurs when surgeons have been able to repair arteries but not the much smaller and thinner-walled veins, leading to build-up of blood pressure in the reconstructed or reattached tissue. This can be relieved by using leeches to drink some of the excess blood and to inject anticlotting agents, giving time for natural healing of the tissue. The technique has been used successfully during surgery to reattach or repair eyelids, ears, penises, fingers, and toes.

Three other groups of worms, now included within the annelids, each used to have phylum status until their probable evolutionary position was resolved by molecular analysis. These are the Echiura, or spoon worms; Sipuncula, or peanut worms; and the deep-sea Pogonophora, or beard worms. The former two groups are not segmented, and it is thought they have lost this character during evolution, in much the same way as leeches, although to a more extreme extent. The pogonophorans were also long thought to be unsegmented, until 1964 when some specimens dredged from the sea floor were found to have a short segmented tail, or 'opisthosoma'. It was suddenly realized that all previous descriptions were – rather embarrassingly – based on incomplete specimens. Many pogonophorans live in muddy burrows and are incredibly thin, elongate animals just a few centimetres in length. Others are giants, however, and build erect tubes stuck to rocks in the depths of the ocean.

The first large tube worms were discovered in 1969 by the United States Navy while operating a deep-sea submersible off the coast of Baja, California. Although these animals, *Lamellibrachia barhami*, were 60 to 70 centimetres in length, and dwarfed all previously known pogonophorans, it was only a few more years before some real giants were found.

The most famous discoveries came in the late 1970s when geologists using the deep-ocean submersible *Alvin* started to

6. Phylum Annelida: A, earthworm; B, ragworm; C, medicinal leech, *Hirudo*; D, pogonophoran, *Riftia*

explore an area of underwater volcanic activity close to the Galapagos Islands. Here were discovered massive chimneys of volcanic rock spewing out hot water rich in toxic chemicals, such as hydrogen sulphide. Astonishingly, life was found to be abundant even in this extreme environment, including forests of giant tube worms up to 1.5 metres in length. These animals, named *Riftia pachyptila*, were adorned with crowns of blood-red tentacles, which made for a striking spectacle from the windows of the deep-sea vehicle. One intriguing fact in common between *Riftia*, *Lamellibrachia*, and all the other pogonophorans, is that something quite central is missing – the gut. They have no mouth, no anus, and no obvious way to eat anything. The clue to how deep-sea tube worms survive is found inside their bodies. Here, a unique organ called the trophosome is packed with millions of living bacteria, all of a type that uses the normally toxic hydrogen sulphide as an energy source to build food chemicals. In the dark depths of the ocean, these bacteria make food by 'chemosynthesis': an analogous reaction to photosynthesis used by plants, but using energy from chemical bonds instead of energy from the Sun. Deep-sea tube worms are farmers, and since their bacterial farm is inside the body, they do not need to eat.

Platyhelminthes and Nemertea: the flat and the slow

Not all worms belong to the phylum Annelida. Flatworms, flukes, and tapeworms are placed together in another phylum, the Platyhelminthes, and quite unlike annelids they have no trace of a fluid-filled cavity (coelom) in their body. Flatworms are rather solid animals that do not wriggle or writhe in the manner of annelids, because their muscle blocks are not split into individual segments and because the lack of a fluid skeleton means there is nothing rigid for the muscles to bend. Instead, flatworms move by sending small ripples of muscular contraction along their edges or, in the smallest species, by using cilia protruding from surface cells. Their lack of a circulating blood system or specialized gills means these animals rely on simple diffusion across the body surface to get oxygen to their cells, which in turn restricts most members of the group to a small size and flattened shape. Flatworms can easily be found by turning over small rocks in streams and rivers, where their oval bodies, a few millimetres to a centimetre in length, creep slowly and steadily along, grazing on algae and debris.

Not all platyhelminths live such innocuous lives, however, and several have unwelcome associations with humans. Perhaps the most important is the fluke *Schistosoma mansoni*, the agent of bilharzia, which currently infects over 200 million people. Symptoms of infection are variable, but in severe cases bilharzia can cause damage to internal organs and even death. Like many flukes, the bilharzia parasite needs two different hosts to complete its life cycle. After developing in a fresh-water snail, the schistosome larvae are released into the river where they will seek out and penetrate the skin of the second host, usually a human.

Another phylum of unsegmented worms is the Nemertea, or ribbon worms, commonly found on the underside of seashore rocks. These worms live life in the slow lane, creeping along at a leisurely pace and generally expending little energy. They too lack large fluid-filled cavities, enabling them to deform and stretch their bodies into contorted shapes, like stringy pieces of chewing gum. Despite their sluggish lifestyle, many ribbon worms are voracious predators that catch and eat other invertebrates using a long proboscis armed with either sticky glue or poisonous barbs. Most ribbon worms are only a few centimetres in length, although one British species, the bootlace worm *Lineus longissimus*, has a serious claim to be the longest animal on the planet. Specimens certainly reach 30 metres, about the same as a blue whale, but claims in excess of 50 metres have been made. Even in these monsters, however, the body of the worm is never more than a few millimetres wide.

7. **A, Phylum Platyhelminthes, flatworm, *Dugesia*; B, Phylum Nemertea, ribbon worm, *Lineus ruber*; C–E, Phylum Mollusca: C, giant squid, *Architeuthis*; D, gastropod, *Murex brandaris*; E, aeolid nudibranch**

Mollusca: from squid to snail

The honour of being the 'largest' invertebrate is usually given to a member of a different phylum: the Mollusca. The giant squid, *Architeuthis*, is an immense animal. It may not grow to anywhere near the length of the bootlace worm, reaching a 'mere' 13 metres, but on sheer bulk it wins hands down, or rather tentacles down. Like other squid, *Architeuthis* has a relatively short stumpy body, eight arms bearing suckers, and two additional tentacles with serrated suckers at the end. The animal lives in the ocean depths and, even though zoologist Tsunemi Kubodera recently succeeded in photographing and filming live giant squid, most of our knowledge still comes from specimens washed up onto land or accidentally caught by fishing trawlers. Alongside the ancient myths and legends of monstrous 'kraken', there is at least one plausible account of a giant squid attacking a ship, logged in the diaries of a Norwegian naval vessel, the *Brunswick*, from the 1930s. The French trimaran *Geronimo* also encountered a giant squid while competing for the Jules Verne Trophy in 2003, with one crew member describing its tentacles as being 'as thick as my arm'. Stories of squid attacking swimmers are more likely to involve a different animal, the Humboldt squid, *Dosidicus gigas*, which grows to a chunky 2 metres. These bold predators hunt in large shoals and can strike at fish, and other swimming prey, with rapidity and ferocity. No wonder some divers have taken to wearing body armour when swimming with packs of Humboldt squid. Along with octopus and cuttlefish, squid are cephalopods – one of the major groups within the Mollusca. Size is not their only claim to fame, and octopus in particular probably have the highest cognitive development of any invertebrate. Their brain is large and complex, their vision is acute, and they are able to solve puzzles such as mazes designed to test spatial memory.

Unlike most cephalopods, the majority of molluscs have a conspicuous shell. Secreted by a specialized layer of cells, the mantle, and composed of bricks of calcium carbonate, the main function of a shell is protection from predators. In gastropod molluscs, such as snails, the single shell is carried on the animal's back, with many of the delicate internal organs hidden away inside. Despite its obvious usefulness, some groups of gastropods have lost the shell completely in evolution. Many of these have found alternative means of protection. Terrestrial slugs, loathed by gardeners, secrete distasteful slime which deters some, though not all, predators. The lack of a shell is a positive advantage in some ways because, unlike snails, slugs can thrive in habitats with low calcium content.

One group of sea slugs, only distantly related to the land-living slugs and snails, has evolved an even more impressive means of defence. Aeolid nudibranchs feed on cnidarians such as sea anemones, but instead of being stung, they manage to harvest the nematocysts, or stinging organelles, without them firing. These subcellular structures are then recycled by the sea slugs and loaded into fronds of tissue growing from the upper surface of their body. Consequently, just like sea anemones and jellyfish, these marine gastropods bristle with a stolen armoury of explosive poison-tipped harpoons. Animals from the third major group of molluscs, the bivalves, have two shells. Oysters, clams, and mussels are well-known examples, and each has a similar way of life. Hidden between the two shells, the animal has elaborate W-shaped gills covered in thousands of cilia which beat to suck in a strong current of fresh, oxygen-rich water. The water flow also carries in tiny particles of suspended food such as microscopic algae that are swept towards the animal's mouth.

Molluscs have been an important food source for humans for millennia. Vast 'middens' or mounds of ancient discarded shells, often hundreds of metres in length, are found throughout coastal regions of the world. In addition to their use as food, both Pliny the Elder and Aristotle wrote of useful pigments that could be obtained from molluscs, particularly the dye Tyrian purple. Used to colour the robes of Greek and Roman noblemen, this vivid dye was extracted from the marine gastropod *Murex brandaris* by heating its bodily extracts after mixing with salt. Some species of mollusc impact on humans in more detrimental ways, such as *Biomphalaria*, the fresh-water snail encountered earlier as the intermediate host for the bilharzia parasite. One mollusc may have even changed the course of European history. In 1588, the

Spanish Armada sailed for England, determined to overthrow the monarch Queen Elizabeth I. The Spanish defeat is well recorded, but perhaps not all the credit should go to Sir Francis Drake. Before sailing for battle, the Spanish fleet anchored in Lisbon harbour for several months, where their wooden hulls became infested with the wood-boring bivalve *Toredo navalis*. This mollusc, the notorious 'shipworm', has an elongated body with its two shells reduced to small plates at one end, used for tunnelling into its food source – wood. With their timbers riddled with holes, the Armada was fatally weakened even before the battle began. The shipworm's habits are also the reason why so few historic sailing vessels remain today. The Swedish battle galleon *Vasa* is a beautifully preserved exception; *Vasa* sank on her maiden voyage in 1628 in the Baltic, a sea not salty enough for *Toredo* to live.

All the above phyla – Annelida, Platyhelminthes, Nemertea, and Mollusca – are part of the Lophotrochozoa: one giant arm of the evolutionary tree of animals with one giant name. The 'trocho' part of the name is derived from the 'trochophore', a particular type of planktonic larva possessed by some, though not all, species in these phyla (most clearly in marine annelids and molluscs). Trochophore larvae are usually described as resembling miniature spinning tops, although since trochophores do not actually spin, perhaps pear-shaped is a more useful description. The 'lopho' part of 'Lophotrochozoa' refers to the 'lophophore', an unusual feeding structure resembling a crown of arms, found in none of the phyla discussed above. Instead, lophophores are found in three additional phyla that are not particularly worm-like. These are the shelled Brachiopoda, the rare Phoronida and the minute Bryozoa, or moss animals, commonly found as mat-like colonies on fronds of large seaweeds. The best way to see a lophophore in action is to find a frond of kelp washed into a rock pool, and examine any white 'sea mats' under a low-power microscope or hand lens. Within a few minutes of being submersed, hundreds of tiny bryozoans will start wafting their delicate tentacles in the sea water, searching for particles of food. It took DNA sequence comparisons to reveal that the animals with lophophores fall in the same part of the Animal Kingdom as the animals with trochophores. This group, the Lophotrochozoa, is the 'sister' superphylum to another great grouping of bilaterian animals: the Ecdysozoa.

Chapter 7
Ecdysozoa: insects and nematodes

To a good approximation, all species are insects.

Robert May, *Nature*, 324 (1986): 514–15

Insects: masters of the land

Nobody knows how many species of insect exist. Estimates range from a few million to over 30 million. At least 800,000 different species have been described and named formally, but even this figure is not known accurately since no master list has ever been compiled. And for those species that have been described, the geographic distribution, ecology, and behaviour are unknown for a large proportion. But why are there so many species of insect? This is not a simple question to answer, but the reasons are likely to include a body plan that can be readily adapted to many ecological niches and food plants, coupled with the great diversity of plant species, particularly in the tropics. In addition, the land-living insects emerged from the sea early in animal evolution, almost 400 million years ago, giving time for tremendous diversification alongside the evolving land plants.

Insects are the supreme land animals. They are part of the phylum Arthropoda and, like all arthropods, insects have an external hard skeleton which is moulted intermittently to allow growth, plus a series of jointed limbs used for locomotion and feeding. Although their ancestors lived in the sea, the insects have evolved a suite of adaptations to permit life in a harsh environment prone to extremes of temperature and severely limited in water – the hostile environment we call land. An external skeleton provides inherent support to the body, whether on sea or on land, but in insects this cuticle has been waterproofed by addition of waxes into the outermost layer. This effectively stops desiccation caused by evaporation from the outer body surface. This solves part of the problem, but still leaves two processes prone to water loss. First, animals need to obtain oxygen and get rid of carbon dioxide, and the physics of gaseous diffusion dictates that this is most efficient across a wet surface. To avoid exposing wet surfaces to the outside world, which would defeat the point of a waterproof exoskeleton, insects have evolved an elaborate series of cuticle-lined 'trachea', or tubes which twist and branch from closeable pores on the outside of the body right into the inner tissues of the animal. Here the cuticle covering is absent and gas exchange can occur exactly where it is needed. Second, all animals need to dispose of nitrogen-containing waste products, which are generated during metabolism of proteins but which can be toxic to cells. Many animals, including humans, overcome this by diluting the waste products and excreting liquid urine, but this is wasteful of water. Insects mainly use a different metabolic pathway and generate uric acid, which crystallizes as a solid – in contrast to soluble ammonia or urea – and then use efficient glands to resorb water before excretion. Since uric acid is not toxic, many insects store some of the chemical in specialized cells, while others actually use it. The white colouration of *Pieris* butterflies, such as the Large White, is generated by uric acid storage in the wing scales.

An obvious characteristic shared by all members of the phylum Arthropoda is segmentation. Most of the major body parts, including muscles and nerves concerned with movement, are repeated in a serial manner along the length of the body, as if the body is divided into a set of units. This type of organization is similar to that of segmented worms, the Annelida, although –

contrary to long-held views – these two phyla are not at all closely related. Annelids fall within the Lophotrochozoa; arthropods are in the Ecdysozoa. In arthropods, the segmentation also affects the rigid external skeleton, so joints of softer cuticle are needed between segments to allow the body to twist and move. Without joints, the animal would be encased in an immovable coat of armour. In insects, the pattern of segmentation has been modified in a consistent way that may partly underpin the remarkable adaptability of insects. Instead of having a series of near-identical segments running the length of the body, groups of segments are fused together into three principal units, or 'tagmata'. First, there is the head, built of six or seven segments fused together, and containing the major nerve concentrations, sense organs, and jointed feeding structures. After a flexible neck joint comes the thorax, built of three segments fused together, each of which has a pair of jointed legs. In most insects, the second and third segments of the thorax, T2 and T3, each also have a pair of wings. And finally, the abdomen, made of eight to eleven segments fused together, albeit less rigidly, has no legs but encases the bulk of the digestive, reproductive, and excretory organs of the animal. In functional terms, the head is concerned with feeding and sensing, the thorax with movement, and the abdomen with metabolism and breeding. This separation of functions has allowed for optimization of each part of the body.

Ruling the skies: wings and flight

Powered flight has evolved only four times in the history of life on Earth – in birds, bats, pterosaurs, and insects. The only flying invertebrates are insects. They are also the most abundant, and the most diverse, fliers on the planet. Flight is absolutely key to understanding the success of the insects. It is interesting, though a little puzzling, that insects evolved two pairs of wings. Are two pairs better than one? After all, birds and bats have only one pair of wings, although some dinosaurs, probably related to the ancestors of birds, had feathers on both 'arms' and 'legs'. The reason for the difference in numbers of wings might relate to a limitation imposed by the mode of embryonic development in vertebrates. There is some evidence that vertebrates, such as bats and birds, are constrained to have only two pairs of limbs; if one is needed for walking, then only one pair is left for flying. The wings of insects, in contrast, did not evolve from legs and no such constraint exists. So while segment T1 just has legs, segments T2 and T3 have both wings *and* legs. Having two pairs of wings allows for even greater diversity in the range of flying styles that insects can adopt.

The insects are divided into almost 30 'orders' including the grasshoppers (Orthoptera), dragonflies (Odonata), mayflies (Ephemeroptera), stick insects (Phasmida), earwigs (Dermaptera), cockroaches (Dictyoptera), land and water bugs (Hemiptera), and fleas (Siphonaptera). But without doubt, the 'big four' of the insect orders, accounting for over 80% of described species, are the beetles (Coleoptera), butterflies and moths (Lepidoptera), bees, wasps, and ants (Hymenoptera), and flies (Diptera). Each is phenomenally diverse, and each has adapted its wings in a different way.

Lepidoptera have two fully developed pairs of wings. In some moths, there are spines or projections that link fore- and hind-wings together, but in many Lepidoptera the wings can be moved and controlled independently in flight. Wing shape varies enormously, from the feathery projections of plume moths to the elongate blade-shaped wings of South American *Heliconius* butterflies and the broad gliding wings of swallowtail butterflies.

Moths and butterflies may look delicate and ephemeral, but some species are robust and long-lived. Monarch butterflies, *Danaus plexippus*, overwinter in mass roosts in central Mexico, before undertaking a collective migration across North America. Each individual will fly for hundreds of kilometres, and within just a few butterfly generations their descendants can reach as far north

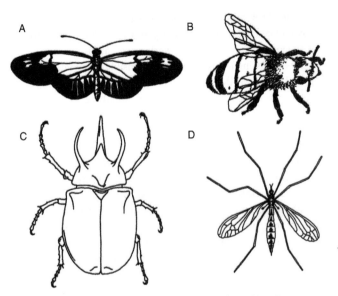

8. Insects, the 'big four': A, Lepidoptera, butterfly, *Heliconius*;
B, Hymenoptera, honeybee, *Apis*; C, Coleoptera, beetle, *Chalcosoma*;
D, Diptera, crane fly, *Tipula*

as Canada – 4000 kilometres from the winter roosts. The Painted
Lady butterfly, *Vanessa cardui*, is also famed for its migratory
behaviour. Few European naturalists will forget the years 1996
and 2009 when vast swarms of Painted Ladies swept north across
Europe from the Atlas Mountains of Africa, breeding as they went,
to eventually reach as far north as Scotland and Finland.

The ants, bees, and wasps – Hymenoptera – also have two pairs of
wings, but these are generally held tightly together by a row of
hooks on the hind-wing. Most species are adapted for rapid,
controlled flight, allowing bees to hover or dart into small spaces
to collect nectar, hornets to catch prey in flight, and parasitic
wasps to land near caterpillars into which they will lay their eggs.
It is primarily within the Hymenoptera that we also see the

evolution of colonies of individuals living together, and even the division of labour. A colony of honeybees, for example, has just one queen but several thousand worker bees, all sisters of the queen bee. Having just one female responsible for laying eggs, while all the others do tasks such as food gathering, cleaning, and defence, is very unusual and takes some explaining. Why should hundreds, or even thousands, of worker bees, ants, or wasps forgo reproduction, while devoting their energies to helping another individual? How could such a situation evolve? The answers are not straightforward. An explanation that was popular for many years was rooted in 'haplodiploidy' – the unusual sex determination mechanism found in Hymenoptera. In many animals, males and females differ because of a single sex chromosome, such as the X and Y chromosomes of humans. But in bees, ants, and wasps, the males have only half the number of chromosomes as found in females. This is because eggs that are fertilized by sperm become females; eggs that remain unfertilized, instead of dying, become males. Under this weird genetic system, sisters – such as worker bees and queen bees – are genetically very similar to each other. Indeed, a female ant, bee, or wasp is more related to her sisters than to her own children. This might imply that cooperation between sisters is favoured evolutionarily, because by helping the queen, the workers are incidentally promoting survival of their own genetic lineage. However, the catch with this often quoted explanation is that, in haplodiploidy, sisters are only weakly related to their brothers, cancelling out the genetic advantage. Instead, the evolutionary origins of sociality in ants, bees, and wasps may have less to do with unusual genetics, and more to do with shared defence of resources by relatives and a breeding system in which extended care of the young is the norm.

The linking together of fore- and hind-wings, as in Hymenoptera, means that two pairs have virtually the same mechanical properties as one pair. Two of the largest orders of insects have gone one step further and only use a single pair of wings for flight. Coleoptera, or beetles, fly using only their hind-wings. Diptera, the true flies, use only their fore-wings. Both groups certainly evolved from insects that used two pairs of wings for flight. In beetles, the ancestral fore-wings have evolved into hardened wing cases (elytra) that cover and protect the hind-wings when not in use. This modification has opened up new ecological niches for beetles; they can burrow into soil, bore into seeds, or tunnel in rotting wood without damaging their thin and delicate flying wings. The great diversity of beetles has fascinated generations of biologists, and the young Charles Darwin was a Coleoptera fanatic. In *The Descent of Man*, Darwin enthuses about one of the giant scarab beetles, writing:

> If we could imagine a male *Chalcosoma* with its polished, bronzed coat of mail, and vast complex horns, magnified to the size of a horse or even of a dog, it would be one of the most imposing animals in the world.

In the true flies, Diptera, the ancestral hind-wings have been modified into tiny club-shaped 'halteres'. These are vibrated up and down during flight, out of phase with the flapping of the fore-wings, and they form part of an intricate sensory feedback system. If the fly's body tilts to one side, the halteres are liable to continue their original plane of vibration, with the inertia of a gyroscope, and sense organs at the base of the haltere will then detect the change in angle between haltere and body position. The fly, therefore, receives continuous information about its precise orientation in space. Consequently, true flies are the most agile of all insects, able to hover, dart, or reverse with the most astonishing rapidity and accuracy.

Amongst the tens of thousands of dipteran species are several that have profound impacts on human life. These include mosquitoes that transmit the malaria parasite or carry the viruses causing yellow fever and dengue fever. Over one million people die from mosquito-borne diseases every year. Other flies have beneficial effects, for example as pollinators, while one species, *Drosophila melanogaster*, has played an important role in medical research. This species of fruitfly has been a favourite 'model organism' in genetics research for over a century, being small and easy to breed in large numbers. Research using *Drosophila* has given deep insights into gene functions and interactions, with relevance to many human conditions including cancer.

Yet more arthropods: spiders, centipedes, and crustaceans

In addition to insects, there are three other classes of living animals within the phylum Arthropoda. Two of these, the chelicerates and myriapods, have successfully invaded land. The third, the crustaceans, is mainly aquatic but with some terrestrial members. Chelicerates include spiders and scorpions and, although most live on land, they had their origins in the sea. Their structure and adaptations to terrestrial life are so different to those of insects that it is clear the chelicerates and insects invaded land independently.

Turning to myriapods, the centipedes and millipedes are the best-known groups. These animals have a specialized head followed by a series of many segments bearing jointed legs. In centipedes, the body segments are separated by flexible rings of cuticle, allowing them to twist, turn, and run swiftly. Centipedes are predators that actively chase and catch their prey, attacking using ferocious 'poison claws' – giant venomous fangs that evolved from the front pair of legs. Millipedes, in contrast, eat mostly wood or decaying leaves, and are much slower animals. They lack poison claws, and in many species the segments interlock to allow millipedes to drive through soil or rotting vegetation like slow-motion battering rams. It is not true that centipedes have a hundred legs, or that millipedes have a thousand, although when it comes to numbers, there are some peculiarities that are not fully understood. Oddly, centipedes always have an uneven number of pairs of walking legs (not counting the poison claws), which means that while centipedes can have as few as 30 legs (= 2 x 15) or as many as 382 (= 2 x 191), no species has exactly 100. Even in species where segment number varies, individuals always differ by a multiple of two pairs. Millipedes have a different oddity. When viewed from above, millipedes seem to have two pairs of legs for each segment, leading to the idea that pairs of segments became stuck together in evolution to form 'diplosegments'.

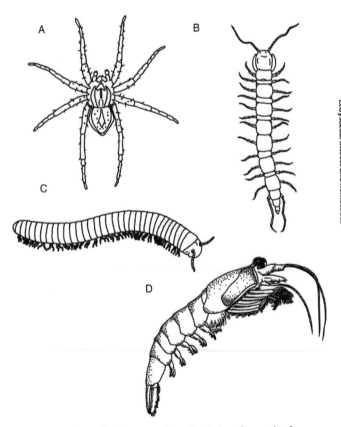

9. Arthropoda: A, Chelicerata, spider; B, Myriapoda, centipede; C, Myriapoda, millipede; D, Crustacea, krill

However, this pattern is not seen from below, and recent gene expression studies have revealed that the boundaries between segments are set independently on the top and bottom sides. In millipedes, segments are no longer simple repeating units.

Like insects, myriapods have trachea to deliver oxygen to their tissues and have limbs that do not branch. The structure of the head is also very similar between centipedes, millipedes, and insects. For over a century, these similarities persuaded biologists that myriapods and insects were close relatives – sisters within the arthropods. Molecular evidence points in a different direction, however, and strongly indicates that insects are actually closer to crustaceans. Indeed, insects probably lie within the crustaceans. Since crustaceans are primarily an aquatic group, the implication is that insects and myriapods invaded land quite separately, and each group evolved adaptations such as trachea and unbranched legs independently to cope with life in their new environment. Crustaceans are a diverse group, including many familiar animals such as crabs, lobsters, and shrimps, and even some parasitic species such as fish lice. Many are hugely important ecologically, such as the copepods found by the billion in marine plankton or the vast shoals of krill on which baleen whales feed. Arguably the most unusual, yet amongst the most familiar, are barnacles, which start their life as free-swimming larvae in the sea before settling onto rocks, sticking down head-first, and spending the rest of their life waving their legs to catch particles of food.

Water bears and velvet worms

Two phyla of animals closely related to arthropods rank amongst the favourite animals of nearly every zoologist. These are the microscopic tardigrades and the forest-dwelling onychophorans. Both types of animal have stubby limbs and soft cuticles, rather than the rigid jointed limbs and tough exoskeletons of arthropods such as insects and spiders. Tardigrades, or 'water bears', are less than a millimetre in length, and can be found living in the surface water on damp moss or lichens.

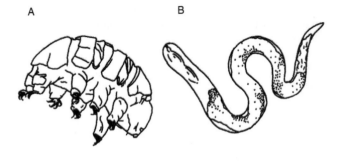

10. A, Tardigrade, or water bear; B, Nematode, or roundworm;
C, Onychophoran, or velvet worm

Watched down a microscope, their chubby bodies and
lolloping gait really do give them the appearance of miniature
bears, albeit bears with eight legs. Apart from their general
cuteness, tardigrades are famous for a remarkable ability to
withstand extreme environmental conditions. If their habitat
slowly dries up, tardigrades secrete a waxy covering and
withdraw their legs until they resemble tiny barrels. They
then vastly reduce their consumption of oxygen and water
until they enter a state of almost suspended animation. In this
condition, known as cryptobiosis (meaning 'hidden life'),
tardigrades can survive for several years. Claims of a century
have been made; however, such extensive survival times are
unlikely in the light of recent studies. Tardigrades are also
remarkably resilient, and once they have entered cryptobiosis,
they have been known to survive temperatures as low as
−200°C or as high as +150°C. Life is put on hold until
conditions improve.

Onychophorans, or 'velvet worms', live on land, and can be found in moist habitats, such as rotten logs and leaf litter, in tropical forests of South America or in the cooler forests of New Zealand. They are soft, slightly 'furry', caterpillar-like animals, a few centimetres in length, with around 20 pairs of short, soft, stumpy legs. Despite their slow-moving nature, most velvet worms are actually hunters, feeding on termites and other insects. Since small insects can move faster than velvet worms, they cannot chase and catch their prey. Instead, they shoot it. Velvet worms have unusual appendages either side of the head, thought to have evolved from legs, which are used to fire streams of sticky slime at their intended targets. This glue entangles the insect prey, which can now be eaten at leisure by the velvet worm. So as not to waste valuable energy, the protein-based glue is also eaten.

Moulting worms

The nematodes, or roundworms, are the most unlikely relatives of the arthropods. They are not segmented, they have no external skeleton, and they have no limbs. As their name suggests, they are simply worms – long, thin, and flexible. Yet, since 1997, more and more DNA sequence evidence has accumulated to indicate that nematodes do indeed sit quite close in the Animal Kingdom to the arthropods, water bears, and velvet worms, plus a few obscure animals such as the wonderfully named (but microscopic) kinorhynchs, or 'mud dragons'. Most zoologists were very surprised by the finding, which had never been suggested from studies of anatomy. But in fact, all these animals have one fundamental character in common – they shed their skins as they grow. Arthropods have a hard external skeleton, which must be shed repeatedly to allow the body beneath to enlarge, a process called ecdysis. Water bears and velvet worms (as well as immature insects such as caterpillars) have softer, more flexible cuticles, but these are also moulted because the molecular structure of their cuticle is not well suited to expansion. Nematode worms have complex cuticles built from tightly wrapped fibres of protein, winding round the body to form layers upon layers of densely packed elastic springs. These too must be shed to allow growth. When DNA evidence indicated that these animal phyla were related, a name had to be invented for the group. Anna Marie Aguinaldo, James Lake, and colleagues, the biologists who first spotted the relationship, named the group Ecdysozoa meaning 'moulting animals'.

Nematodes have a very unusual internal structure. They have a fluid-filled space in their body, as do many other worms, but they keep this fluid at very high pressure, around ten times the pressure of fluid in other worms. The internal pressure pushes out against the nematode's tissues and cuticle, giving them a circular cross-section. This explains their common name of 'roundworms'. Another peculiarity of nematodes is that all their body muscles run in the head-to-tail orientation (longitudinal), with no muscles wrapping around the body in circles. Most other worms, including earthworms, ragworms, and ribbon worms, have both types of muscle, which can contract in opposition to each other, or antagonize, allowing the animal to change shape and crawl or burrow. So how do nematodes manage to wriggle and move, if their muscles can only contract lengthways? The answer lies in the high-pressure fluid cavity and the elastic springy cuticle, which counteract the muscles, enabling the worm to throw its body into rapid undulating waves. Their movement is not so well coordinated as in earthworms or ragworms, partly because of the unusual muscle arrangement, but also because roundworms are not segmented and cannot easily make different parts of the body move in opposite directions. Instead, nematodes move with a thrashing motion, which is not very effective in swimming but works perfectly well in their preferred home – inside things. Many species of nematode live in soil or rotting vegetation; rotting fruit can be swarming with them. There is even a yeast-eating 'beermat nematode', *Panagrellus redivivus*. Many others are parasitic inside plants or other animals. Nor are humans immune from their attention, with some serious medical conditions caused by parasitic nematodes, including river blindness, guinea worm disease, toxocariasis, and elephantiasis.

The propensity of nematodes to live inside other organisms is described in a poetic, but somewhat exaggerated, 1914 quote from the 'father of nematology', Nathan Augustus Cobb:

> If all the matter in the universe except the nematodes were swept away, our world would still be dimly recognizable, and if, as disembodied spirits, we could then investigate it, we should find its mountains, hills, vales, rivers, lakes and oceans represented by a thin film of nematodes. The location of towns would be decipherable, since for every massing of human beings there would be a corresponding massing of certain nematodes.

Closely related to the nematodes, and with many similarities, are a group of exceedingly long and thin worms given their own phylum, the Nematomorpha. Although rarely more than a millimetre in width, these animals often reach 50 to 100 centimetres in length. Like nematodes, the nematomorphs have a tough cuticle, which is moulted as they grow, and only longitudinal muscle. Unlike nematodes, they do not eat anything. Or at least the adults do not, and the gut is shrivelled to a mere remnant. Juvenile nematomorphs certainly do feed, eating from the inside the body tissues of their arthropod host, which may be a grasshopper, cockroach, or freshwater shrimp. There, the worm will grow and moult, ever lengthening, until it reaches a size too great for the host animal, whereupon it bursts or crawls out, leaving behind the carcass of the unfortunate host. The adult worms must live in water, and so if the host is a land-living animal such as a cockroach, the parasite somehow manipulates the behaviour of the host, persuading it to move towards water and dive in to await its grisly death. The nematomorphs that parasitize water-living hosts, such as freshwater shrimps, gave rise to the common name for these animals: 'horsehair worms'. Long before the true life cycle of these animals was known, country people would occasionally notice long, thin worms swimming in apparently clean horses' drinking troughs, when no such creatures were present the day before. The myth arose that these were hairs from horses' tails that had come to life. The truth is less miraculous but more macabre: the giant worms were parasites that had burst out of tiny shrimps living unobtrusively in the water.

Chapter 8
Deuterostomes I: starfish, sea squirts, and amphioxus

> I also here salute the echinoderms as a noble group especially
> designed to puzzle the zoologist.
>
> Libbie Hyman, *The Invertebrates IV* (1955)

Clues from embryos

Echinoderms have been called the strangest animals on Earth –
and with good reason. Starfish, sea urchins, brittlestars, sea
cucumbers, and sea lilies – the five classes within the echinoderm
phylum – have much in common with each other, but little with
anything else. They are built like nothing else on the planet. Even
so, it has long been realized that they sit in the same part of the
Animal Kingdom as do you and I: the deuterostomes. And the
first clue to that relationship is to be found in embryos.

Most animals begin their life as a single cell, the fertilized egg.
This divides to give two cells, then four cells, then eight, then
sixteen, and so on. Although this seems straightforward, several
different patterns can be seen if different bilaterian animals are
compared. Two of the commonest are spiral cleavage and radial
cleavage. The differences between them are quite clear if
developing embryos are watched down a microscope. In spiral
cleavage, when four cells divide to make eight, the new cells end
up sitting above the grooves between the four old cells. If you tried

to stack four oranges on top of four other oranges, this is exactly the pattern you would opt for. But in radial cleavage, the new cells sit directly on top of the four old cells, in a manner that would demand considerable balancing skills if attempted with oranges.

Whether spiral or radial, the same pattern is repeated in each subsequent cell division until eventually a hollow ball of cells is formed. Starting at a point or slit on the ball's surface, some cells then move inwards, rather like a finger or hand being pushed into an inflated balloon. The indentation where the cell sheet folds inwards is called the blastopore, the original ball of cells being the blastula. As the embryo develops further, this indented tube eventually forms the gut. And it is during this process that a second significant difference is found between the two patterns. In animals with spiral cleavage, the blastopore may mark the 'mouth' end of the gut, or more usually the blastopore is slit-shaped and closes up in the middle to leave two open ends: mouth and anus. But in animals with radial cleavage, the blastopore marks the rear end of the embryo, where just the anus will form. The mouth must break through quite separately at the other end of the developing embryo, as the incipient gut tunnels through to the far side. For this reason, animals with spiral cleavage have long been called 'protostomes', meaning 'first mouth' and alluding to the observation that the mouth develops from the first opening to form in the embryo. Animals with radial cleavage and blastopore at the rear were 'deuterostomes', meaning 'second mouth'. Inevitably, not all animals fit into either of these neat patterns, particularly if their embryos are endowed with lots of yolk which affects how the cells divide.

The distinctions were made by Karl Grobben in 1908, but a century later they need to be used with caution. Two of the great superphyla of bilaterian animals defined by molecular analyses, the Lophotrochozoa and the Ecdysozoa, include all the animals with the protostome mode of development, but they also include many other animals. For example, ecdysozoans such as insects and nematodes do not have spiral cleavage, nor indeed is it radial. Even so, 'protostome' is still often used today as a term to mean the Lophotrochozoa plus Ecdysozoa, with the knowledge that it is just a handy name not a unifying rule. Just as confusingly, the evolutionary grouping now called the Deuterostomia, and again defined by molecular analyses, is slightly different from Grobben's original. The group as now defined includes only some, but not all, animals with radial cleavage and secondary mouth formation. Perhaps it would have been better to discard the old names, but nomenclature is not always logical. Instead, simply note that not all protostomes have protostomous development, and indeed some animals placed in the protostomes actually have deuterostomous development. Consequently, not all animals with deuterostomous development are 'true' deuterostomes. There are only three major animal phyla in the Deuterostomia, as defined today, plus possibly one or two 'minor' phyla. The three major groups are the Echinodermata, the Hemichordata, and the Chordata.

Life with the number five

Cut an apple open across its middle, and you will see a five-pointed star containing seeds. Take a close look at a wild rose and notice its five petals. Whether looking at fruit, flowers, or even leaf patterns, the number five is pervasive across the Plant Kingdom, and is the basis for much variation and adaptation. Animals, in contrast, despise the number five. One might point out that we have five fingers, but since we have two hands, the true number, of course, is actually ten (or twenty, if we count all digits). Five does not lend itself well to animals that have a central plane of symmetry, with a left- and right-hand side, as seen throughout most of the Animal Kingdom. Numbers such as two, four, six, and eight are everywhere, but not five. The basal animals, such as jellyfish in the phylum Cnidaria, do not have an obvious plane of left–right symmetry, but even these animals usually have four-fold symmetry, not five-fold.

Echinoderms do things differently. The evolution of the whole phylum has been dominated by the number five. The pattern is most easily seen in the starfish and brittlestars, common invertebrates of the sea shore and subtidal zone, which have five arms radiating from a central region or disk. In starfish, also called sea stars, the arms are held out quite rigidly. When the animal moves, it seems to glide over the sea bed, a trick achieved by the use of thousands of tiny 'tube-feet' projecting from the underside. The movement of the tube-feet is driven by an extensive series of fluid-filled canals in the body, unique to echinoderms, called the water vascular system. Although superficially similar to starfish, the brittlestars are different in that their five arms are thinner and more flexible, and used to assist the animal's locomotion by grasping and pulling. The two groups are also quite different ecologically, particularly from the viewpoint of a scallop. Brittlestars graze on debris and detritus, ingesting tiny pieces through a small downward-pointing mouth located in the middle of the central disk. Most starfish, on the other hand, are voracious predators. They may move slowly, but this does not matter if your prey doesn't move at all. Many starfish hunt bivalve molluscs such as mussels, oysters, and clams, animals that live sedentary lives inside their two tight-fitting shells. While generally protected from predators, starfish are the bivalves' worst nightmare. When a hunting starfish encounters prey such as a clam, it wraps its arms around it, holding tight using the sucker-like tube-feet, and pulls. As a tiny gap appears between the shells, the starfish then (rather riskily) pushes part of its own stomach inside-out, through its own mouth, and presses it into the gap. The stomach secretes enzymes to break down protein, weakening the clam's own muscles, so that the shells can be prised further apart. Eventually, the body of the clam is exposed and devoured. No wonder that those few species of bivalve that can swim, such as Queen scallops, make a break for freedom at the merest smell of a starfish.

The sea urchins, which bristle with defensive spines, and the soft, elongated sea cucumbers, are also echinoderms. Here, the

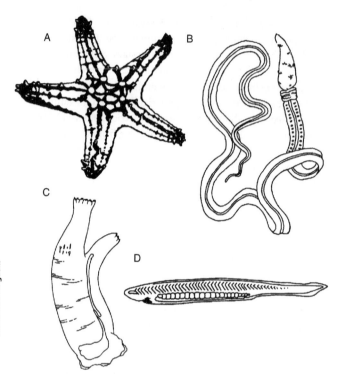

11. A, Phylum Echinodermata, starfish; B, Phylum Hemichordata, acorn worm; C–D, Phylum Chordata: C, adult ascidian, *Ciona*; D, amphioxus, *Branchiostoma*

number five is less obvious to a casual glance, but it is certainly there. In each case, there are five zones around the body bearing tube-feet: an indication that these animals evolved from starfish-like ancestors in which the arms were folded up over the rest of the body. The fifth group in the phylum, crinoids or sea lilies, comprises filter-feeding animals with a mouth on the upper side, surrounded by a crown of five feathery arms. This sometimes sits atop a stalk, particularly in deep-sea species. With their tube-feet and upward-facing mouth, crinoids are oriented upside-down in comparison to starfish and brittlestars.

The evolutionary origin of five-fold, or pentaradial, symmetry in echinoderms is intriguing. It is clear that pentaradial symmetry evolved from bilateral (left–right) symmetry, for three key reasons. First, echinoderm larvae are bilateral, just like the larvae of many other marine animals. It is only when they settle out of the plankton and undergo metamorphosis that the five-fold pattern emerges. Second, echinoderm fossils have been found with all manner of symmetries, including bilateral, suggesting that the five-fold pattern was hit upon rather late in their evolution. Third, and most important, in the evolutionary tree of animals, the phylum Echinodermata is nested well within the Bilateria, indicating evolution from the same common ancestor as shared with all the bilateral animals alive today.

Hemichordates: stinking worms

Some years ago, I wanted to collect an acorn worm. These are rather curious, non-segmented worms belonging to the Hemichordata, a phylum close to echinoderms on the evolutionary tree. Their early embryonic development is very similar to that of echinoderms, with radial cleavage and 'secondary' mouth formation, and their larvae – found occasionally in plankton samples – could easily be confused for echinoderm larvae. I had never seen an acorn worm in the wild, and I needed a sample for a research project. But on writing to marine biologists, I received one reply that baffled me. The writer told me that he too had never seen one in Britain, but he was absolutely certain they existed on a particular beach because he was sure he had smelt one. Of course, I was never going to believe the evidence of a simple aroma! That is, until I collected them myself. Most hemichordates are just a few centimetres in length, and live hidden away in sandy or muddy burrows where they filter food particles from the overlying sea water. They do this using a system of particle-trapping slits in their throat (pharyngeal slits) across which water passes, in a similar way to water passing over the gills of a fish. Indeed, the system is probably homologous, meaning that the shared ancestor of hemichordates and vertebrates had slits in the throat, used for acquiring food or oxygen. Many acorn worms do indeed have a striking medicinal smell, rather like iodine, which has been traced to a toxic chemical – 2,6-dibromophenol – found at high concentrations in their skin. The function of the substance is not fully clear, but it may deter predators intent on devouring them or it may limit bacterial growth in the burrow, or indeed both. Regardless of its adaptive function, the smell lingers on clothes and fingers, and once encountered is never forgotten.

The acorn worms are not the only members of the phylum Hemichordata. Their evolutionary sisters are a group of animals called pterobranchs: minuscule tube-dwelling animals with a crown of tentacles. They are not often encountered unless you know exactly where to look. The best-known British species, *Rhabdopleura compacta*, is less than a millimetre in length, and its tiny white tubes are found predominantly on the inner surface of the discarded shells of one species of mollusc, the dog cockle *Glycymeris glycymeris* – and even then, only in a few selected locations around the British coast. Other species can be found in Bermuda and in the fjords of Scandinavia, but their biology is still not well known. A distinct pterobranch genus, *Cephalodiscus*, was discovered on the seabed of the Magellan Straits by the famous *HMS Challenger* expedition in 1876, and when it was realized that this animal had pharyngeal slits, it became evident that pterobranchs were related to acorn worms, in the Hemichordata. A third pterobranch genus, *Atubaria*, is unusual in that it does not seem to live in a tube. Little is known about the biology of *Atubaria* because just 43 specimens have ever been seen, these all having been collected on 19 August 1935 by a marine expedition from the Imperial Palace of Japan.

Tunicates: was man once a leather bottle?

Pull up the mooring ropes attached to buoys in a sea-water harbour and you will probably find them encrusted with hundreds of bottle-shaped leathery lumps, often in shades of yellow or brown, each a few centimetres in length. Pull one off its submerged home and it may squirt you in the eye. These animals, though they hardly look like animals, are sea squirts, or ascidians. Despite their amorphous appearance, these lumps are among our evolutionary cousins, members of our own phylum: the Chordata. Externally, sea squirts are encased in a tough outer covering or tunic which to the touch feels more plant-like than animal-like. This is because, remarkably, the tunic contains cellulose, a chemical usually found in plants not animals. There are two tubes, or 'siphons', on the top of the body, and sea water is sucked in through one and expelled through the other, the current being driven by the wafting of thousands of tiny cilia within the animal. This constant flow of water brings microscopic food particles and dissolved oxygen, and removes waste products.

A phylum is supposed to comprise a group of evolutionarily related animals with similar body layout. To repeat Valentine's words, 'phyla are morphologically-based branches of the tree of life'. So how could a sea squirt be in the same phylum as the vertebrates, along with you and me, along with birds and fishes? Looking at an adult sea squirt – the stationary, filter-feeding, cellulose-coated lump – there is little to suggest a close evolutionary relationship. And indeed, the earliest naturalists were completely unaware of the link. Aristotle considered sea squirts to be molluscs, like clams and snails, but he did note they were unusual in that their 'shell', actually the tunic, was leathery rather than hard and enveloped the whole animal. In the early 19th century, Lamarck removed them from the molluscs and erected a new group, the Tunicata, but he did not solve their affinities. All changed in 1866 when the brilliant Russian zoologist

Alexander Kowalevsky published a careful description of embryonic and larval development of a sea squirt, and moreover realized the deep significance of what he had found. The embryos of sea squirts develop into miniature 'tadpoles', usually just a millimetre in length, that swim in the sea for a day or two before landing head-down on a rock or other substrate. There they undergo a dramatic metamorphosis into a miniature version of the adult. From that moment on, the animal never moves place again and remains stuck to its landing spot, filtering sea water. Kowalevsky found that in the swimming tadpole stage, there is a small brain at the front, connected to a nerve cord along the back, which in turn lies on top of a stiffening rod, the notochord. These were all features characteristic of vertebrates, such as humans and fish, or at least characteristic of their embryos. An evolutionary relationship to vertebrates was clear.

News of the discovery swept through scientific circles, since there had already been much debate about which groups of invertebrates might be closely related to the vertebrates. In *The Descent of Man*, published in 1871, Darwin wrote:

> Some observations lately made by M. Kowalevsky, since confirmed by Prof. Kuppfer, will form a discovery of extraordinary interest.... The discovery is that the larvae of Ascidians are related to the Vertebrata, in their manner of development, in the relative position of the nervous system, and in possessing a structure closely like the chorda dorsalis of vertebrate animals. It thus appears, if we may rely on embryology, which has always proved the safest guide in classification, that we have at last gained a clue to the source whence the Vertebrata have been derived.

The emerging view, shared by Darwin, was that the long-extinct common ancestor from which sea squirts and vertebrates both evolved must have been a small tadpole-like animal, with the various features seen today in a sea squirt larva. But several other zoologists attempted to derive vertebrates from an ancestor much more like a modern sea squirt, complete with metamorphosis, and this view pervaded until late in the 20th century. Charles Neaves, a Victorian lawyer and poet who wrote much about evolution, beer, and women's rights, espoused this latter view in rhyme:

> How many wondrous things there be, of which we can't the reason see!
> And this is one I used to think, that most men like a drop of drink.
> But here comes Darwin with his plan, and shows the true Descent of Man:
> And that explains it all full well, for man-was-once – a leather bottèl!

Neaves may have found himself a reason to enjoy a drink (which he expanded on in eight subsequent verses), but he was not accurately reflecting Darwin's view or Kowalevsky's discovery. There is no need to assume that the common ancestor of sea squirts and vertebrates, our distant ancestor, had a life cycle that underwent metamorphosis like a modern sea squirt. Indeed, there are living relatives of sea squirts, called larvaceans, which never undergo the change and remain as swimming tadpoles all their life, feeding and reproducing.

Amphioxus: the riddle of the sands

The phylum Chordata, or chordates, can be split into three evolutionary groups, or subphyla. In addition to the tunicates (such as sea squirts and larvaceans) and the vertebrates, there is a fascinating group of marine chordates called cephalochordates, more usually referred to as amphioxus or lancelets. Several characteristic features typify all chordates. These are a brain, a nerve cord running along the back rather than the belly, a notochord, repeated blocks of muscle on each side of the body, and 'pharyngeal' slits or holes between the throat and the outside world. These characters describe the typical body plan of a chordate. In the sea squirts, we saw most of these characters in the tadpole larva, except for the pharyngeal slits; it is the adult sea squirt that has these filtering structures. All the chordate features are seen in fish, with gills developing from pharyngeal slits, but the notochord becomes surrounded by bone during development and rather squeezed out of existence. As humans, we also have most of these characters at some stage in our development, but again the notochord is really only evident in the embryo, while our pharyngeal slits are just grooves in the embryo that never break through as actual holes. But the animal in which all the chordate features are most evident, even in the adult, is amphioxus. It has the clearest example of the chordate body plan one could ever hope to see.

There are about 30 species of amphioxus found in marine habitats around the world, often in tropical or subtropical seas, but sometimes in colder waters. One species lives off the European coast, and can be found buried in gravel in parts of the Mediterranean Sea and in the English Channel close to the treacherous Eddystone Reef with its famous lighthouse. Another species is common in subtidal sands around the Gulf Coast of Florida, while a third species was once so abundant near the city of Xiamen in China that there used to be a commercial fishery harvesting the animals for food. All species are roughly fish-like in overall appearance, just a few centimetres in length, with segmentally repeated muscles on either side of a prominent notochord acting as a stiffening rod. The springy notochord acts as an antagonist to the contracting muscles; this allows the animals to swim very rapidly when needed, for example when emerging from the sands to shed eggs or sperm into the sea water. The pharyngeal slits are very obvious and used for filtering algae from a current of sea water drawn in through the mouth. Unlike true fish, which are vertebrates, there is no bone, no fins emerging from the sides of the body, and a far less complex head. Amphioxus has the basic 'chassis' of a chordate, without many of the complications that have evolved in vertebrates.

A century ago, amphioxus was one of the most popular topics of research in all of zoology. In 1911, the great German evolutionary biologist Ernst Haeckel wrote that amphioxus was 'after man the most important and interesting of all animals'. I am tempted to agree with him. Even so, for Haeckel and his contemporaries, it posed a conundrum. On the one hand, many zoologists thought amphioxus must be a degenerate vertebrate, simply a fish that had lost many specialized characters. Others thought this highly unlikely, with Edwin Stephen Goodrich, Britain's greatest comparative anatomist, calling such a suggestion 'ridiculous'. Goodrich's view, backed up by careful studies of the animal's development and anatomy, was that amphioxus retained a more primitive chordate organization, not so different from the long-extinct ancestral chordate. This opinion was eventually accepted, and recently gained even stronger support from genome sequencing. Amphioxus, therefore, is another crucial link between invertebrates and vertebrates, rather like the position accorded to the sea squirt larva. In amphioxus, we have an animal, still living today, that has most vertebrate features in rudimentary form. It does have some of its own specializations, not least in its strange one-eyed head, akin to the cyclops of Greek mythology. These small changes occurred during the half a billion years since amphioxus and vertebrates last shared a common ancestor; exactly the same time frame that saw the emergence and diversification of the fish, amphibians, reptiles, birds, and mammals. So while amphioxus is not an ancestor of any other living animal, it seems to have changed remarkably little from the long-extinct ancestor of all the vertebrates.

Chapter 9

Deuterostomes II: the rise of vertebrates

> The Romans in the height of their glory have made fish the
> mistress of all their entertainments; they have had Musick to
> usher in their Sturgeons, Lampreys and Mullets.
>
> Isaak Walton, *The Compleat Angler* (1653)

The great divide?

It is common to see textbooks of zoology concerned solely with the
invertebrate animals, and other books devoted to the vertebrates
(animals with backbones). Many university courses divide
teaching of animal diversity along the same lines. And it is not a
new division. Jean-Baptiste Lamarck, chiefly remembered today
for his discredited ideas on the inheritance of acquired characters,
made the distinction clear and wrote of '*des animaux sans
vertebres*' two hundred years ago. Even he was not the first to draw
this line, since over two thousand years ago, Aristotle divided the
animals into 'sanguineous' (those with blood) and 'non-
sanguineous' (those without); this was essentially the same
vertebrate/invertebrate divide.

Despite its persistence and popularity, many zoologists have
pointed to a deep-rooted problem with this view. The majority of
individual animals are invertebrates, and most of the described
animal species are also invertebrates. The difference in numbers is

vast, with millions of invertebrate species and only around 50,000 different vertebrates. But the problem runs deeper than simply inequality. The problem lies in the evolutionary tree of animals, the history of animal life. Animals are classified into phyla, which represent branches on the evolutionary tree containing species with similar body organization. Of the 33 or so animal phyla, 32 are purely invertebrate. Even the 33rd is not a phylum solely of vertebrates, but contains a mix of invertebrate and vertebrate animals. It is our phylum, of course, the Chordata, and contains the invertebrate tunicates, the invertebrate amphioxus, and the vertebrates. The organization of the body is sufficiently similar in all these animals that they are grouped together. So taking a step back and looking at the diversity of the Animal Kingdom, vertebrates are not considered sufficiently different to even warrant their own phylum. Does this mean that vertebrates are but a twig in the tree of animal life?

Differences run deep

While the numbers argument and the phylogenetic problem cannot be doubted, vertebrates do have some important specializations. Indeed, in some ways, they are exceptional animals. Most obviously, almost all the 'large' animals on the planet are vertebrates. There are a few big invertebrates, such as squid, octopus, and Goliath beetles, but the vast majority of invertebrates are less than a few centimetres in length. Contrast that to vertebrates, where large size is the rule. Fish, amphibians, reptiles, birds, and mammals – all have their giants. The 12-metre whale shark, 1.5-metre giant salamanders, 30-metre dinosaurs (extinct of course), the 3-metre-tall elephant bird (also sadly extinct), and the 30-metre blue whale may be record-holders, but each is simply at the end of a continuum. In fact, very small size is almost unheard of amongst the vertebrates. One of the most diminutive is an Indonesian fish, *Paedocypris*, with adults under a centimetre. Even this is a monster compared to many invertebrates.

One key to growing larger is rooted in the sophisticated system of veins and arteries used to deliver oxygen and remove carbon dioxide from active tissues deep inside the body: the highly efficient 'closed blood system'. Incidentally, some of the largest invertebrates, squid and octopus, also have a closed circulatory system, although this evolved independently. A second and equally important character permitting large size is the one that gives the vertebrates their name: the vertebral column. Skeletons come in many forms around the Animal Kingdom. Many worms have fluid-based support systems, arthropods have tough external skeletons, and echinoderms have rigid internal plates of calcium carbonate. But the skeleton of vertebrates is different and quite remarkable. In some vertebrates, the skeleton is made of cartilage, a tough yet flexible protein-based tissue, but in most it is bone. Not only is bone surprisingly light and strong, making it very effective at supporting bulky bodies, but it has an extra and unexpected quality: it is alive. Intermixed in a matrix of protein and minerals are cells that deposit bone and others that remove bone. Other cells sense mechanical pressures and relay messages instructing the bone to grow or shrink and respond to the changing conditions. Bone is always dynamic. It is an extraordinary tissue and one ideally suited to large, active, growing animals, whether living in water or on land.

Vertebrates also differ from their closest invertebrate relatives, the tunicates and amphioxus, in their sophisticated brain and sense organs. The organization of the brain is very consistent across all the vertebrates, from lampreys to humans, with the three key sensory inputs being visual (paired eyes), chemical (paired olfactory sense organs), and mechanical (detecting pressure changes in water or sound in air). The entire head region of vertebrates is elaborate, based around a skull that encases the brain, yet presenting these sense organs to the outside world. The embryonic development of the skull reveals another oddity: a peculiar type of cell called the neural crest. These cells arise from the edges of the developing nerve cord and migrate through the embryonic tissues, before forming a whole variety of structures, including the bone or cartilage of the skull, jaws, and gill supports. Without neural crest cells, vertebrates could not build their complex, protected head region; without neural crest cells, they could never be the large predators and herbivores that dominate ecosystems on land and in the sea.

These features – large size, efficient blood circulation, dynamic skeleton, intricate brain, protective skull, and elaborate sense organs – combine to set vertebrates apart from their relatives. They may share the chordate phylum with amphioxus and tunicates, but the vertebrate body is much more complex and sophisticated. The differences within the phylum may run even deeper. Comparisons between the genome sequences from vertebrate and invertebrate animals have uncovered a fascinating fact. DNA sequences clearly show that early in vertebrate evolution, right at the base or very soon after, a major mutation occurred. The whole genome – every single gene – was doubled, and then doubled again. For every gene in an ancestral chordate, the early vertebrates had up to four. Some of these 'extra' genes were soon lost, but many were not, and consequently vertebrates have a greater diversity of genes than do most invertebrates. Whether new genes permitted the evolution of new vertebrate features is controversial, but one thing is certain: the invertebrate/ vertebrate divide should not be ignored.

The vertebrate tree

A common way to classify backboned animals is to divide them into fish, amphibians, reptiles, birds, and mammals. This works reasonably well for many purposes, but it does not accurately reflect the phylogenetic tree of vertebrates. One problem is that the various species of fish do not sit together on one single evolutionary line, separate from the other groups. Animals called 'fish' are found mixed in with other vertebrates. A similar problem applies to 'reptiles' since living reptiles share an evolutionary lineage with the birds. If we are to be strict in classifying animals according to evolutionary history, then groups called fish and reptiles should not exist.

Despite this complication, the path of vertebrate evolution, as known from fossils, molecular biology, and anatomy, is quite simple. The first vertebrates to evolve were fish-like in shape but lacked biting jaws. Although quite diverse in their heyday, over 400 million years ago, there are now only two surviving lineages of these jawless wonders: the lampreys and hagfish. The vertebrates with jaws evolved from jawless ancestors, and these early predators diversified into three main evolutionary lineages. These three groups are the 'chondrichthyans' (including sharks with their cartilaginous skeletons), the 'actinopterygians' (ray-finned fish), and 'sarcopterygians' (lobe-finned vertebrates). All three lineages include aquatic 'fish', but the sarcopterygians also include the vertebrates that left the water and emerged onto land. These animals, with strong skeletal structures in their fleshy fins, became the 'tetrapods' – vertebrates with four limbs. They comprise the amphibians, the 'reptiles', the birds (which perch on the same evolutionary twig as some reptiles), and the mammals.

Lampreys and hagfish: surfeits and slime

With no biting jaws, the lampreys and hagfish need other ways to get food into their mouths. Adult lampreys have a sucker-like cup enclosing a circular rasping mouth armed with rings of sharp teeth. This ferocious-looking apparatus enables the animal to attach itself firmly to the flesh of its living prey, usually a large fish, and to suck its blood. Lampreys can stay fastened to their prey for several weeks, hanging limply as parasitic appendages.

When lampreys are attached by their sucker to prey, or even just to stones on a river bed, they cannot get oxygen from water taken in through the mouth. Instead, lampreys have 'tidal' gills: water is drawn into holes on the side of the animal's head, and expelled

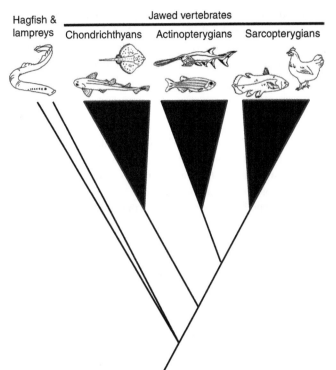

Hagfish & lampreys

Jawed vertebrates

Chondrichthyans Actinopterygians Sarcopterygians

12. Phylogenetic tree of vertebrates

from the same. Larval lampreys, or ammocoetes, in contrast, have a normal unidirectional flow of water – in through the mouth, over the gills, and out through the gill slits – something made possible because juvenile lampreys are not parasitic. Lampreys spawn in shallow gravelly rivers, and after hatching the developing ammocoetes burrow into thick mud where they stay for several years, living on particles of food extracted from decaying matter. These slippery worm-shaped larvae can easily be found by digging in deep mud, whenever it is found close to fast shallows, in many British streams and rivers. After metamorphosis, when the sucker is formed, the adults of most species migrate to the sea. There are, however, several land-locked species, such as the diminutive (and non-parasitic) brook lamprey *Lampetra planeri*, which grows to just 15 centimetres, compared to the 1-metre-long sea lamprey *Petromyzon marinus*.

Lampreys have a long association with the monarchy. King Henry I, son of William the Conqueror, died after eating 'a surfeit of lampreys', his favourite dish, when visiting his grandchildren in Normandy in 1135. Undaunted, his grandson Henry II also indulged in the jawless treat, and Henry III had a regular supply of lamprey pies baked for him 'since after lamprey all fish seem insipid'. Continuing the royal tradition, the City of Gloucester sent lamprey pies for Queen Victoria's Diamond Jubilee in 1897 and for Queen Elizabeth's Silver Jubilee in 1977.

Hagfish are similar to lampreys in lacking jaws, but instead of a sucker, they have tentacles surrounding two sideways biting plates and a horny, retractable tongue. They prey on living invertebrates, such as marine worms, and will also rasp at the flesh of dead and dying fish on the sea floor. Hagfish will even gnaw their way into the dead bodies of larger animals, including whales and large fish, and eat them from the inside. Like lampreys, hagfish lack 'paired fins' located on the sides of the body; paired pectoral and pelvic fins are features characteristic of living jawed fish and, now transformed into legs, their land-dwelling descendants. The hagfish vertebral column is rudimentary compared even to that of lampreys, and for this reason, some zoologists do not even call hagfish 'vertebrates', using the term 'craniate' to encompass hagfish, lampreys, and jawed vertebrates. However, this view is contentious because features can be lost in evolution, and this might well be the case with hagfish vertebrae. We should not remove an animal from its natural group for reasons of secondarily losing something. In terms of overall anatomy and pattern of embryonic development, hagfish are really very similar to the rest of the vertebrates.

They do have a few peculiarities, however, and none more spectacular than slime. Many animals are slimy, but hagfish take it to a new level. Hagfish are the unchallenged masters of slime. When a hagfish is disturbed, pores along the sides of its body start to release a protein-based secretion that expands massively on contact with water. The quantity is impressive. Within seconds, a small 20-centimetre hagfish can produce several handfuls of thick, gluey slime, well suited to deterring predators. To avoid getting tangled in its own slime, the hagfish has a neat trick. It ties itself into a simple overhand knot which it then slides along its own body, wiping it clean.

Jaws: sharks, skates, and rays

Everyone knows that sharks have jaws, whether fans of the 1975 movie or not. Jaws, together with paired fins, are the defining features of the three largest evolutionary groups of living vertebrates – the chondrichthyans (such as sharks and dogfish), actinopterygians, and sarcopterygians. In the words of Alfred Sherwood Romer, 'perhaps the greatest of all advances in vertebrate history was the development of jaws'. Studies on the embryos of dogfish, and other jawed vertebrates, have clearly shown how these highly efficient feeding structures evolved. In embryonic development, streams of migratory neural crest cells move down from the edges of the forming hindbrain into a series of bulges where they construct the skeletal supports for gills. In the jawed vertebrates – from sharks to humans – one of these bulges, the mandibular arch, develops not into a gill support, but into the bones or cartilage of the jaws. The stream of cells just behind it, the hyoid arch, forms a support structure connecting the back of the jaws to the skull. These pathways and patterns of cell migration in the embryo reveal that jaws almost certainly evolved from modified gill supports.

In living sharks, the upper jaw is not fused to the skull above but hangs quite loosely from elastic ligaments plus the hyoid-derived

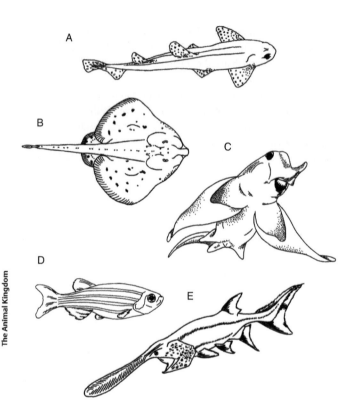

13. A–C, Chondrichthyans: A, dogfish; B, little skate; C, chimera;
D–E, Actinopterygians: D, zebrafish; E, paddlefish

support at the back. This allows a feeding shark to protrude both
jaws, either to pick small prey items delicately off the sea floor or
to sink its teeth deeply into the flesh of large prey. The teeth of
most sharks are sharply serrated and efficient at slicing through
tissue once the jaws have been pressed deep into the prey, the
cutting helped by the shark thrashing its body from side to side.
Finding prey involves an impressive and sophisticated suite of
sense organs. Sharks have a sense of smell that is both sensitive
and directional. In some sharks, notably the Hammerhead

(*Sphyrna*), the two nostrils are placed far apart on strange extensions from the sides of the head, allowing them to deduce even more precisely the direction of highest chemical concentration. As they swim closer to their prey, sharks use both vision and mechanical sensing of vibrations in the water to home in. Just as they strike, many sharks slide a protective membrane over their eyes to prevent damage, which immediately reduces their vision. Temporary blindness does not give a chance for the prey to escape, however, as the shark then relies on exquisitely sensitive electroreceptors to detect the weak electric fields produced by animal muscles. These sensory cells are found in the 'ampullae of Lorenzini', a series of specialized pits in the shark's skin. They were first described in 1678 by Stefano Lorenzini, a brilliant Italian anatomist later imprisoned by the Grand Duke of Tuscany because of an alleged friendship with the Duke's estranged wife.

One notable character that sets sharks, dogfish, skates, and rays apart from most other jawed vertebrates is that their skeletons are made of cartilage not bone. They also lack a gas-filled cavity in their body; such a structure, called a swim bladder, is found in ray-finned fish and evolved into the lungs of land vertebrates. Without a swim bladder, it might be thought that sharks would sink to the bottom of the sea if they stopped swimming, but this is not the case. Instead, sharks solve the buoyancy problem in a quite different way. The key adaptation is a giant liver packed with oil, especially the long-chain hydrocarbon squalene, the low density of which counteracts the high density of the shark's skeleton, teeth, and scales, making sharks neutrally buoyant. Stability and additional lift is provided by stout paired fins on either side of the body. Buoyancy is also not a problem for most skates and rays, close relatives of the sharks, since these animals are usually benthic, meaning they live on the sea floor, in contrast to the pelagic, or free-swimming, sharks. Some giant rays, such as the manta ray, spend less time on the bottom and instead cruise through the ocean, flapping their greatly enlarged pectoral fins and straining plankton through a mesh of spongy tissue attached to their gill arches.

The final group of chondrichthyans, evolutionarily quite divergent from the sharks, dogfish, skates, and rays, are the strange ratfish, or chimeras. They too have a cartilaginous skeleton and lack a swim bladder, and they also have internal fertilization like sharks. But chimeras differ from other cartilaginous fish in that the upper jaw is fused to the skull and there is only a single gill opening on each side instead of several. The solidly built head with a fleshy elephant-like snout gives these animals a freakish appearance. They have an overall 'fishy' shape with large fins, but their big eyes and buckteeth best resemble a cartoon rabbit, while some species also have a long, rat-like tail. Appropriately, the name 'chimera' recalls the mythological monster of ancient Greece, made from components of various animals, described by Homer in the *Iliad* as 'a thing of immortal make, not human, lion-fronted and snake behind, a goat in the middle'.

The ray-finned fish: flexibility

Most of the well-known species of fish belong to a diverse group called the actinopterygians, or ray-finned fish. Examples include commercial fish such as cod, haddock, herring, tuna, and eels; species kept in aquaria such as goldfish, tetras, guppies, and catfish; most fish sought by anglers such as trout, carp, pike, roach, and bass; and many others, including minnows, sticklebacks, and gobies. It is easier to list the 'fish' that are not actinopterygians, since these are only the hagfish, lampreys, sharks, skates, rays, chimeras, coelacanths, and lungfish. There are over 24,000 species of ray-finned fish inhabiting the world's oceans, seas, rivers, and lakes.

Ray-finned fish have both 'unpaired' and 'paired' fins, just like sharks. The unpaired fins lie along the midline of the body, and comprise one or more dorsal fins along the back, a caudal fin at the tail, and an anal fin on the ventral side. There are also the two sets of paired fins: pectorals just behind the gills and pelvics further back. In ray-finned fish, as the name suggests, the fins are supported by thin bony rays, giving them great manoeuvrability. This is particularly important for the pectoral fins, which can be twisted and flexed, enabling fine control whether the fish is swimming, turning, or even remaining motionless in the water. During the evolution of the actinopterygians, fins have been modified in many different ways, and this was clearly one factor underlying the diversification of this group. To give some extreme examples, knifefish can swim slowly either forwards or backwards using ripples sent down an enormously enlarged anal fin, while flying fish have large wing-like pectoral fins enabling them to glide through air for up to 50 metres. Tuna fish out-sprint their prey with bursts of speed made possible by concentrating movement to the caudal fin and posterior body; sea horses lack a caudal fin entirely and swim slowly using undulations of the dorsal fin.

The bony rays supporting the fins also show variation. In some species, they serve a defensive function, such as the sharp and protruding spines in the dorsal fin of a perch or stickleback, and in a few cases, they can inject venom, for example in stonefish, weever fish, and lionfish. They can be also used as accessory organs for feeding, such as in gurnards and angler fish. Gurnards are bottom-living fish with elongated pectoral fin rays, equipped with a range of sensory receptors, used for 'walking' along the sea bed and feeling for prey. In angler fish, the first three spines of the dorsal fin are exceptionally long and fused to form a 'fishing rod' used to lure prey towards a gaping mouth.

Ray-finned fish have solved the problem of density in a different way to sharks. Just below the vertebral column is a gas-filled cavity, the swim bladder, which acts as an internal float to help maintain buoyancy. In some fish, such as carp and trout, the swim bladder is connected by a tube to the gut, allowing it to be filled by gulping air at the surface. In other fish, such as perch, the connection to the gut has been lost and the swim bladder is filled by a specialized gland which secretes gases absorbed from the blood. Many fresh-water fish, including members of the carp family, also use their swim bladder to enhance hearing, using modified spines on the vertebrae to transmit vibrations of the gas bladder to the inner ear. Perhaps surprisingly, some fish can also make sounds with their swim bladder, used for attracting a mate or deterring rivals. For example, male toadfish (family Batrachoididae) make sounds by contracting fast 'sonic muscles' attached to the swim bladder, causing its walls to vibrate rapidly. This produces a sound like a loud, plaintive foghorn.

The head of ray-finned fish is complex and intricate. In most ray-finned fish, the left and right lower jaws can swing sideways, while one of the upper jaw bones, the premaxilla, can protrude forwards. These movements allow the mouth cavity to be enlarged suddenly, creating a strong suction force, used to capture prey items that would otherwise escape. Suction feeding is seen in a huge number of ray-finned fish and underpins the ecology of many species. At the back of the head are the gills. These are hidden behind a flap, the operculum, which not only protects the delicate gills but also plays a key role in how they function. By closing the operculum and widening the mouth, then shutting the mouth and opening the operculum, ray-finned fish effectively pump water across their gills, even when not swimming. Coupled with a blood supply to the gill filaments oriented in the opposite direction to the water flow, this allows ray-finned fish to extract maximal oxygen from water.

The great diversity of ray-finned fish is explained only partly by the invention of the swim bladder, the adaptability of fin rays, suction feeding, and the operculum. High species numbers are dependent on the integration of many factors, including a combination of ecological opportunities, an adaptable body plan, and possibly even features of the genome. On the last point, it is interesting that the teleost fish, which constitute the majority of ray-finned fish, share an additional genome duplication on top of the two that occurred at the base of vertebrates. It is currently unclear if this permitted greater adaptability of the body plan, or if it even caused enhanced speciation rates through different genes being lost in different populations. The extra genome duplication did not affect all ray-finned fish, and there are still a few living descendants from the earliest evolutionary radiation of this group: the so-called 'non-teleost actinopterygians'. These include the peculiar filter-feeding paddlefish (*Polyodon*) with its spatula-shaped head, the heavily armoured gars (*Lepisosteus*), and the various species of sturgeon, many now rare and endangered, whose eggs are prized as caviar.

Chapter 10

Deuterostomes III: vertebrates on land

Eye of newt, and toe of frog,
Wool of bat, and tongue of dog,
Adder's fork, and blind-worm's sting,
Lizard's leg, and howlet's wing,
For a charm of powerful trouble,
Like a hell-broth boil and bubble.

William Shakespeare, *Macbeth*, Act IV, Scene I

From lobe-fins to legs

On 22 December 1938, a young museum curator in South Africa was shown an unusual, iridescent blue fish among the catch of a local fishing boat. The fish, almost 2 metres in length, with strong fleshy fins and heavily armoured scales, became a *cause célèbre*. It was the first living specimen described of a coelacanth, a member of a group of ancient fish with a fossil record dating from 400 million years ago until their supposed extinction 65 million years ago. The *London Illustrated News* described the discovery as 'one of the most amazing events in the realm of natural history in the twentieth century'. Specimens of *Latimeria chalumnae*, named after museum curator Marjorie Courtenay-Latimer, have since been caught many times off the East African coast, particularly near the Comoro Islands, and a second coelacanth species, *Latimeria menadoensis*, has been discovered in the Indian Ocean.

The excitement about living coelacanths is not simply that they were once thought extinct. More importantly, these are animals with special significance for understanding the evolution of land-dwelling vertebrates, a critical step in our own evolutionary history. The fleshy fins, which can be moved independently on the left and right sides as if the coelacanth is 'walking' in the open sea, are central to the argument. Their structure, together with various skull features, reveal that coelacanths belong to the 'sarcopterygian' group (lobe-finned vertebrates), and not to the ray-finned fish. In addition to coelacanths, the other two groups of living lobe-finned vertebrates are the lungfish and the tetrapods, the latter including all the land vertebrates including humans. Neither coelacanths nor lungfish are the actual ancestors of land vertebrates, but the three groups are related and they each descend from lobe-finned fish that swam in the early Devonian period, around 400 million years ago. Fossil evidence and molecular data suggest that tetrapods are evolutionarily slightly closer to lungfish than to coelacanths, but both groups of lobe-finned fish are important for understanding our origins. The living lungfish, of which there are four species in Africa, one in South America, and one in Australia, are all quite unusual and specialized animals, but they are indeed air-breathing fish – they have lungs equivalent to the lungs of land vertebrates.

While several groups of invertebrates, such as insects, myriapods, spiders, and snails, made the difficult transition from living in water to living on land, the same transition was only successful once in the entire evolution of vertebrates. The single evolutionary lineage of vertebrates that overcame the challenges of living on dry land gave rise to all of the land vertebrates still living today: all amphibians, all reptiles, all birds, and all mammals. To live successfully on land, animals must be able to obtain oxygen from air, find and catch food on land, carry their body weight in a medium far less supportive than water, propel themselves along land, and avoid drying out through excessive loss of moisture. Lungfish, close relatives of land vertebrates, can use their lungs to breathe air, as well as their gills to get oxygen from water, suggesting that air-breathing evolved long before the true transition to life on land. But supporting the body, feeding, and moving on land seem more of a challenge, and they demand several evolutionary changes between 'fish' and 'tetrapods'. Some remarkable fossils have cast light on these modifications, and have even revealed the order in which they took place.

One anatomical change was the evolution of a flattened snout capable of snapping at prey, rather than using the suction method that works so well underwater. Fossils of the extinct *Panderichthys* and *Tiktaalik*, which lived around 375 million years ago, show this feature well and must have given these animals a somewhat crocodile-like front end. However, they were still very 'fishy' in that their fins had delicate rays at the end of skeletal elements, rather than strong, bony fingers. This combination of fish-like and tetrapod-like features led Neil Shubin, the discoverer of *Tiktaalik*, to nickname this animal the 'fishapod'. *Acanthostega*, which lived a little later at 365 million years ago, had fins that ended in jointed finger-like elements, making it even more tetrapod-like. Interestingly, there were not just five fingers, as seen today in most living land vertebrates, but eight on the forelimb and probably the same number on the hindlimb. *Acanthostega* almost certainly lived in water and breathed using gills, but it may have been capable of forays onto land, perhaps to catch food or to bask in the sun. Another fossilized early tetrapod, *Ichthyostega*, may represent yet another step in the transition to living on land, since in addition to the above features, this animal had a stronger 'axial' skeleton or backbone, with longer bony projections, or 'zygapophyses', enabling vertebrae to interlock and help support the animal's body weight.

Frogs and salamanders: the skin-breathers

Long-extinct animals from the Devonian period may represent the first tentative forays of backboned animals onto the land, but these animals would have still been heavily dependent on water, not least for reproduction. The latter also holds true for some tetrapods living today, a group of animals that spend much of their life on land, but lay their eggs in or around water. These are the frogs and toads, newts and salamanders, and the legless caecilians: the living amphibians. Most of these animals never stray far from damp habitats, since their skin is not particularly waterproof and, in many species, must be kept moist as a surface for gas exchange. A second reason for reliance on water is that their eggs and young demand a wet habitat. The larvae of most living amphibians, such as the tadpoles of frogs, even have external gills to extract oxygen directly from their aquatic habitat. When discussing amphibians, it is tempting – but inaccurate – to consider the living species as representing just another small stepping stone on the route to 'true' life on land, less successful and less advanced than the reptiles, birds, and mammals. But the fact that they exist at all today is evidence of their continuing success, and in reality, the living amphibians are greatly specialized and very different from the earliest land vertebrates. Furthermore, some individual species are quite numerous, particularly several species of frogs and toads. For example, the Cane Toad *Rhinella marinus* has spread and become so abundant in northern Australia, following its deliberate but disastrous introduction in 1935, that it is now a major invasive pest.

A few species of amphibian spend their entire lives in water, never venturing onto land at all, even as adults. Examples include the Japanese giant salamander *Andrias japonicas*, which can grow to 1.5 metres, the grotesque American 'Hellbender' salamander *Cryptobranchus alleganiensis*, and the African clawed toad, or *Xenopus*. But perhaps the best-known 'aquatic amphibian' is the Mexican axolotl, *Amblystoma mexicanum*, which resembles a large, 20-centimetre-long, sexually mature tadpole, complete with feathery external gills. That is exactly what it is, since axolotls evolved from 'normal' terrestrial salamanders through a change in their developmental physiology, and now become mature without

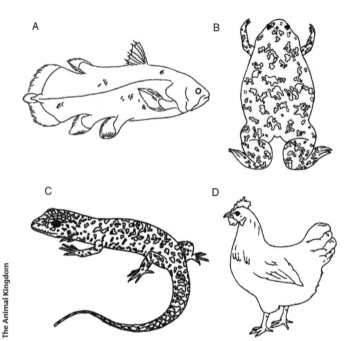

14. Sarcopterygians: A, coelacanth; B, African clawed toad; C, Tasmanian snow skink; D, chicken

going through metamorphosis into the ancestral adult form. The axolotl is a powerful reminder that evolution is not a one-way street and that different lineages of animals each adapt to their local conditions, irrespective of any overarching trends that we may perceive.

Scales and sex: the reptiles

The reptiles represent a 'grade' of organization, rather than a single group on the evolutionary tree of vertebrates, and the living species comprise a disparate assemblage of animals including lizards, snakes, turtles, crocodiles, and the archaic tuatara of New Zealand.

Dinosaurs were also reptiles, on the same evolutionary line as crocodiles and birds, while other extinct reptiles included the winged pterosaurs and the marine ichthyosaurs, mosasaurs, and plesiosaurs. These aquatic species, like the marine turtles of today, went back to water secondarily – they evolved from ancestral species that lived fully on land. The defining feature of the earliest reptiles is that they marked a true break from aquatic habitats: the first group of vertebrates to do so. The reptiles that inhabit land can live, feed, and breed without ever returning to water.

Two key innovations seem to have been central to this transition – the evolution of a thoroughly waterproof skin and the invention of a shelled egg with several internal membranes. The first property seems clear enough, and was achieved by the evolution of more complex skin with several layers of cells producing keratin proteins and lipids. One implication of this change is that the skin cannot be used for respiration (as it is in modern frogs and salamanders) because only wet surfaces allow oxygen and carbon dioxide to diffuse across. Instead, reptiles evolved 'costal respiration' in which muscles attached to the ribs are used to ventilate the lungs, turning the lungs into even more effective breathing organs. The importance of the 'amniotic egg' is less obvious, but was also vital. The key lies in three membranes, the amnion, allantois, and chorion, which envelop the embryo and provide a large expanse of blood vessels for gas exchange, plus a site where toxic nitrogenous waste can be accumulated safely away from the developing body. Although most reptile species, including turtles and crocodiles, lay amniotic eggs encased within a shell, some snakes and lizards have live birth. Most commonly, as in garter snakes, boas, and vipers, this involves the mother retaining the large yolky eggs inside her body throughout their development. Some other reptiles provide nourishment directly from the mother, rather than from yolk; in the most extreme cases, this can involve a placenta, as in the *Mabuya* and *Pseudemoia* skinks. The many physiological, anatomical, and behavioural adaptations of reptiles have allowed them to invade some of the hottest and driest environments on the planet, including the baking deserts of Africa, Australasia, Asia, and the Americas.

The body physiology of reptiles is well suited to warm conditions, since most bask in the sun's rays to increase their body temperature. This enables a high metabolic rate and active lifestyle to be sustained even without good insulation of the body. Temperature also affects the biology of many reptiles in a very different and rather unusual way: it can determine the sex of their offspring. For example, if the eggs of an American alligator are incubated below 30°C, they hatch into females, while eggs kept at 33°C develop as males. This phenomenon, called 'temperature-dependent sex determination', or TSD, contrasts with the more familiar 'genotypic sex determination' in which genetic differences control the sex of offspring, such as the male-determining genes found on the Y chromosome of mammals. But why should some reptiles (and, for that matter, some fish) use TSD, when at first sight the gene-based method seems more reliable? Is there not a danger that a shift in environmental conditions, such as a changing climate, could drive a TSD population to extinction, since every offspring might then develop as the same sex? The answer seems to lie in adaptation to local ecological conditions, as demonstrated neatly by recent research on the Tasmanian snow skink, *Niveoscincus ocellatus*, undertaken by Ido Pen, Tobias Uller, and colleagues. This reptile lives from sea level up to mountainous regions, and – remarkably – populations at low altitude have TSD, while animals of the same species at high altitude use a genotypic method. The reason for the difference seems to be that mothers at low altitude use TSD to produce proportionately more daughters in warm years which then have maximal opportunity to grow large and fecund during the longer summers, but they switch to producing more sons in cooler years because male size is less important in snow skinks. This advantage is lost at high altitudes where growth rates are slower and where the more wildly fluctuating temperatures would play havoc with the ratio between the sexes unless TSD gave way to genotypic sex determination.

Feathers and flight: the birds

One celebrated group of reptiles, the dinosaurs, dominated terrestrial life on Earth for many millions of years. The first dinosaurs evolved around 230 million years ago and diversified into a multitude of species of different sizes, shapes, and habits, until their sudden and famed extinction 65 million years ago. Or to be more accurate, until their apparent extinction. The popular notion of a total wipe-out of dinosaurs is a little deceptive because some animals alive today are direct evolutionary descendants of a group of dinosaurs, the theropods. Well-known extinct theropods include the giant carnivore *Tyrannosaurus* and the smaller, but perhaps equally fearsome, *Velociraptor*, made famous in the film *Jurassic Park*. Of course, *Tyrannosaurus* and *Velociraptor* are no longer roaming the Earth, but you see some of their close relatives every day. One group of theropods did not go extinct 65 million years ago, but survived the global catastrophe and diversified until the present day. They are the birds.

From an evolutionary perspective, birds are a group of dinosaurs that did not become extinct. The idea that birds evolved from dinosaurs was first proposed by Thomas Henry Huxley in the 1860s. Huxley noted key similarities between the layout of the skeleton in theropod dinosaurs and in an extinct bird, *Archaeopteryx*, known from a few beautifully preserved 150-million-year-old fossils. Although *Archaeopteryx* had features that were very lizard-like, such as teeth and a long bony tail, it also had wings and feathers. Then, as now, *Archaeopteryx* was considered to be one of the first birds to evolve. Huxley's views were contentious, and although every biologist accepted that birds evolved from ancient reptiles, the idea that they were actually direct descendants of dinosaurs soon fell out of favour. The idea lay dormant for most of the 20th century until thrust back into the mainstream in the 1970s through the careful work of John Ostrom at Yale University. But the most dramatic and clinching piece of evidence did not come until the 1990s, when several remarkable fossils of 'feathered dinosaurs' were discovered in China – these were undisputed 'non-flying' dinosaurs, but with feathers covering their body and legs. Not only do the feathered dinosaurs provide strong evidence for a bird–dinosaur relationship, they also highlight feathers as an early adaptation, possibly for keeping warm, that paved the way for the later origin of flight.

The feathers of modern birds are remarkable structures. Feathers used in flight have an intricate and asymmetrical structure that provides rigidity and power on the downstroke, while also being strong and incredibly light. They have a central shaft, or rachis, from which protrude a myriad closely spaced barbs, each bearing minute hooked barbules that interlock. In contrast, the 'down feathers' used for body insulation do not interlock in the same way, and so trap pockets of air rather than producing sheet-like surfaces. As well as their two principal functions – flight and insulation against cold – feathers play roles in water-proofing, camouflage, and communication between individuals. Feathers and flight dominate the entire ecology and behaviour of birds, and have combined to shape their evolution. Weight is an important issue in flight, and accordingly birds have evolved to have quite thin, hollow bones, strengthened by internal struts. Heavy teeth have been lost in evolution, as has the long tail. But more important than absolute weight is its distribution, and so the anatomy of birds is adapted to place the centre of gravity further forward than in most vertebrates, directly between the wings. This has been achieved by tucking the 'thighs' of the hind legs forward along the sides of the body and lengthening the foot; this explains why birds' knees seem to point backwards – they are not actually knees but ankles.

There are around 10,000 species of bird alive today, found living on every continent and flying over every sea. They include tiny hummingbirds in South America, spectacular birds of paradise in the forests of New Guinea, majestic condors soaring through Andean passes, shearwaters skimming over ocean waves hundreds of miles from land, kestrels hovering over grassy verges, secretive wrens, robins, thrushes, and more. This may sound like a picture of diversity, but in reality all birds are very much alike – at least in terms of anatomy. The most striking exceptions are the few birds that have secondarily lost the ability to fly. Penguins with their unusual body form are adapted to water not air, and ostriches with their great size and bulk are flightless: these exceptions serve to remind us that flight places massive constraints on the anatomy and physiology of birds. Evolution cannot circumvent the laws of physics.

Milk and hair: the mammals

The active lifestyle of birds is only possible because of their relatively warm body temperature, generated by a high metabolic rate coupled with the insulation provided by feathers. The other land vertebrates that generate and maintain their own body heat are the group to which we belong: the mammals. In the case of mammals, however, the vital insulation is provided by hair. The structure of hair is far less complex than that of feathers, comprising rather simple strands made from fibres of the protein alpha-keratin. Even so, overlapping layers of hair can trap air very effectively and so keep warmth in. This retention of body heat allows mammals to get out and about in cold conditions, before the sun's rays have had a chance to warm their reptilian relatives. Unlike birds, mammals did not evolve from within the familiar diversity of reptiles. In an evolutionary tree of amniotes (the land vertebrates with an amniotic egg), one division gave rise to lizards, snakes, crocodiles, dinosaurs, and birds, while a sister lineage – the extinct synapsids – gave rise to the mammals.

Besides hair, a second important character shared by all mammals is lactation: the production of milk to provision offspring. This is a crucially important adaptation since it allows mammals to reproduce at any time of year, even when a diversity of food is not easily found or its availability fluctuates in time. Adult females can stock up on food whenever it is available and store energy as fat reserves; offspring are then provisioned with high-energy milk by suckling from their mother. Food-gathering by an experienced adult is also likely to be more efficient than by a juvenile, meaning that suckling on milk allows the infant to put a greater proportion of energy towards growth.

Strange as it may sound, the use of milk may have also paved the way for the great ecological diversity of mammals, through the evolution of complex teeth. The argument goes as follows. Because of lactation, newborn mammals do not need teeth. This means that the skull and jaw grow substantially before teeth arise; in turn, this enables a shift away from continual replacement of simple teeth, the system seen in most amniotes including lizards. Instead, mammals evolved 'diphyodonty', meaning that just two sets of teeth are produced: one simple set in juveniles and then complex teeth in the almost full-size jaw. Because tooth development is delayed in this way, mammalian teeth could evolve to have precise matches between the upper and lower jaws, a feature known as occlusion. Such a property is hard to envisage in an animal whose jaw grows extensively while containing teeth. Occlusion gave mammals the crucial ability to chew and grind food, especially tough plant matter, or to slice meat cleanly off their prey. Equipped with such formidable apparatus, the early mammals diversified to exploit a greater range of food sources and feeding strategies than seen in any other group of vertebrates.

There are about 4,400 species of mammals, less than half the number of birds, yet they display a greater diversity of body shapes, sizes, and modes of life. Of these, there are just five living species of monotreme, or egg-laying, mammal: the platypus, and four types of echidna or spiny anteater. All other mammals are 'therians' and have live birth. These include a few hundred species of marsupials which give birth to very immature infants and nourish them within a pouch, such as kangaroos, wombats, opossums, potoroos, bandicoots, koala, and the Tasmanian devil. The vast majority of living mammals are placentals, which have longer pregnancies and no pouch. The ecological diversity of placentals is staggering, and includes insectivores such as shrews, grazing herbivores like antelope, elephants, giraffe, and bison, hunting predators such as foxes and lions, opportunist omnivores such as mice, rats, and humans, aquatic herbivores such as manatees, aquatic predators like seals and dolphins, and even a group of mammals that took to the skies, the bats.

For most of the 20th century, there has been confusion over the true phylogeny of placental mammals. Amongst all this diversity, who is most closely related to whom? The question is now much closer to resolution, particularly since the recent application of DNA sequencing technologies. The emerging consensus divides the placental mammals into four great lineages. Remarkably, these lineages map beautifully onto the known geological history of the continents, suggesting that diversification of the placentals occurred just as the major landmasses of the world were separating. There is the 'Afrotheria', which, as the name suggests, comprises the mammalian orders that originated in Africa, including elephants, aardvark, and manatees. From the Americas came the 'Xenarthra', comprising anteaters, sloths, and armadillos. The Laurasiatheria includes a range of mammals thought to have evolved on the northern supercontinent of Laurasia, the forerunner of Europe plus much of Asia. These include cats, dogs, whales, bats, shrews, cows, and horses, amongst many others. Finally, the Euarchontoglires or Supraprimates includes rats, mice, and rabbits, plus primates such as monkeys and apes.

Standing back and looking at our own place in the evolutionary tree of animals, humans represent but a tiny twig. We sit within the primates, which in turn are within the Euarchontoglires.

These fit within the placentals, which are part of the therians, which lie nested inside the mammals, which are part of the amniotes, in turn within the tetrapods, and then inside the sarcopterygians. The sarcopterygians are one of the three groups of jawed vertebrates, within the vertebrates, inside the chordates, within the deuterostomes, within the bilaterians, inside the great tree of animal evolution.

Chapter 11
Enigmatic animals

There are known knowns; there are things we know that we know. There are known unknowns; that is to say there are things that we now know we don't know. But there are also unknown unknowns, there are things we do not know we don't know.

Donald H. Rumsfeld, US Department
of Defense briefing, 2002

New phyla, new insights

The history of zoology has been a story of changing opinions. There has been a century of debate and argument surrounding the evolutionary relationships between animals, and the problem is compounded by the fact that new species are discovered daily. For hundreds of species, more is learned each year about anatomy, ecology, development, and behaviour. It is pertinent, therefore, to stand back and ask how accurate is our current state of knowledge? Will there be further great overhauls, or do we now have a secure framework from which to delve deeper into animal biology? We must first ask whether we really know the full diversity of the Animal Kingdom.

It is a certainty that many thousands, or even millions, of animal species remain to be discovered. Tropical rainforests and the deep sea are just two ecosystems that teem with life, yet where scientific exploration has just scratched the surface. However, finding a new species, or even a thousand new species, does not radically change our understanding of animal biology. It is certainly significant in other ways; for example, knowing all species in an ecosystem can help in attempts to understand nutrient cycling and patterns of energy flow. Such insights are important. But most new species that are discovered are close relatives of already known species and so, if we wish to comprehend the overall pattern of animal diversity on the planet, such findings are not the key. They add detail, but they do not force a radical change to our state of knowledge.

The story is different at higher taxonomic levels. The most fundamental category in the classification of animals is, of course, the phylum. To use Valentine's words again, 'phyla are morphologically-based branches of the tree of life'. Discovery of a new phylum, therefore, really does change our state of knowledge. It adds a new branch to the animal evolutionary tree, and just as importantly reveals a new morphology: a new way to build the body. Putting the two together – a new branch and a new morphology – in turn can change our views on when, why, and how particular characters arose in evolution: perhaps fundamental characters such as symmetry, segmentation, or a central nervous system. But are there any phyla left to be discovered?

In this book, I recognize 33 different animal phyla, and most of these have been known for a very long time. By the late 20th century, many zoologists thought that all phyla must have been discovered. There was a surprise, then, in 1983 when the Danish zoologist Reinhardt Kristensen described a new species that was so different from everything else that it needed an entirely new phylum. He named this phylum the Loricifera. These minute animals, usually far less than a millimetre in length, look like miniature urns or ice-cream cones, clinging to sand grains. A few other zoologists had noticed these animals in the 1970s, including Robert Higgins, after whom the loriciferans' swimming Higgins larva is now named. But the biggest surprise is that the new phylum was not found in a remote and inaccessible part of the world, but just off the coast of Roscoff in France, home to a busy marine biology research centre.

The next 'new phylum' to be discovered – the Cycliophora – was again simply being overlooked. Cycliophorans are tiny symbiotic animals that live on the mouthparts of scampi (*Nephrops*) and lobsters (*Homarus*). These host species are so common that thousands of people must have eaten a rogue cycliophoran without ever knowing what zoological curiosity they were consuming. Remarkably, it was Kristensen again – together with Peter Funch – who described the new animal in 1995, after it had been first spotted by Tom Fenchel.

A third new body plan was reported in 2000, and this time it really was from a remote field site, visited by very few scientists. It had been found when Kristensen (again) was leading a field trip of students to Disko Island off the coast of Greenland. There, the students discovered some unusual microscopic animals living in an icy freshwater spring. Just one-tenth to one-eighth of a millimetre in length, with complex jaws that could be projected out of the mouth, their anatomy was so different from anything else that they deserved at least a new class, and possibly a new phylum. They were named the Micrognathozoa, meaning 'miniature jawed animals'.

So could there be yet another phylum waiting to be discovered? Quite possibly. The three examples above are all minute animals, far less than a millimetre in length, and it is amongst such microscopic fauna that a similar future discovery might be made. A promising place to look would probably be amongst the 'meiofauna', the animals that live between grains of sand. The discovery of the Loricifera off the French coast implies that such a finding could be made anywhere in the world, yet even so, I would suggest that remote deep-sea habitats might be the best bet. But if you really want to describe a new phylum of animals, I would advise that you don't start with hunting for a new species. Instead, a new phylum might well be present amongst the old.

New phyla from old?

It may sound paradoxical, but many changes to the list of known animal phyla, whether they are discoveries of new phyla or mergers between old phyla, have come about because of more detailed studies on previously described species. A phylum should contain animals from one evolutionary branch; hence, if new data indicate that a phylum contains superficially similar species from different parts of the evolutionary tree, then that phylum must be split in two. There is no alternative. This has happened several times in the past two decades, particularly as DNA sequence data have been used to test the evolutionary relationships between animals. When the DNA data highlight a glaring oddity – a species out of place in the tree – then classification has to change. A new phylum can be made for an old species.

The most important, yet still controversial, examples concern two groups of unusual worms. These are the Acoelomorpha and Xenoturbellida. Although neither contains very familiar animals, or even common animals, they have been known to science for a long time. So while the species are not new, they may still merit one or two new phyla. The Acoelomorpha contains small, marine, flatworm-like creatures, usually a few millimetres in length. One of the easiest to find is the beautiful *Symsagittifera roscoffensis*, or 'mint sauce worm', which is bright green in colour because of algae living inside its body. The worm lives on sandy beaches around Europe, especially on the French coast near Roscoff, where it can be found as 'slicks' of greenish sludge in wet puddles. If you creep slowly towards the sludge, it has the disconcerting habit of disappearing; this is a living sludge made of thousands of green worms that simply crawl into the sand when disturbed. These worms, and many like them, were traditionally placed in the phylum Platyhelminthes, alongside the 'true' flatworms, flukes, and tapeworms. There were always a few dissenting voices calling attention to unusual features of their anatomy, but for the most part, the mint sauce worm and its allies stayed within the Platyhelminthes. Only when gene sequences were compared did it become abundantly clear that they were not at all closely related to flatworms, flukes, and tapeworms, and a new phylum was proposed.

It was a similar story for the Xenoturbellida. These animals are much larger than the acoelomorphs, with the first species *Xenoturbella bocki* from a Swedish fjord being several centimetres in length and a recently discovered Pacific species even bigger. They are also flattened worms and not very impressive to look at, being rather simple, yellowish-brown animals with few discernible organs apart from a blind-ending gut. They too were considered to be platyhelminths by most zoologists, although some argued for closer relationships to echinoderms or hemichordates. One suggestion, based on initial DNA sequence analyses, was that *Xenoturbella* was a mollusc, but this conclusion was shown to be an unfortunate error caused by extracting DNA from *Xenoturbella*'s last meal rather than from its own cells. When enough genuine *Xenoturbella* DNA was extracted, and many gene sequences analysed, it became clear that the animal is not a platyhelminth, nor a mollusc, echinoderm, or hemichordate, but something quite distinct from other animal groups. A new phylum was proposed in 2006.

It seems possible that a few more animal phyla might be found lurking as already known animals, misclassified into the wrong part of animal phylogeny. So where might one look? There are dozens of unusual invertebrate animals that share only some characters with their supposed relatives. The challenge for zoologists is to determine which of these represent simply aberrant members of their phylum – where evolution has modified the body plan – and which ones have been misleading zoologists for decades. For example, Katrine Worsaae has drawn attention to the unusual marine worm *Diurodrilus*, presently considered an annelid but possessing few of the normal annelid characters and possibly even lacking segmentation. *Lobatocerebrum* is a similar case, as this worm has characters of both annelids and platyhelminths. Myzostomids, unusual annelid-like parasites of sea lilies, are another problem group. Might any of these be a new phylum?

Another oddity, and possibly the most bizarre animal on the planet, is *Polypodium hydriforme*. This tiny animal spends most of its life actually inside the eggs of sturgeon, better known as caviar, and when it emerges, it breaks up into a swarm of microscopic jellyfish. It may actually be related to jellyfish and a member of the phylum Cnidaria, but if so, it is certainly a strange one. It might be related to *Buddenbrockia plumatellae*, a weird worm-shaped parasite without a clear front, back, top, bottom, left, or right, and with no central nervous system. Both animals possess structures rather like the stinging capsules of Cnidaria. Molecular analyses suggest that *Buddenbrockia* does indeed belong to the Cnidaria, which means that the phylum to which it was formerly assigned, Myxozoa, must be subsumed within Cnidaria. New data can therefore remove phyla from the list, as well as generate new ones.

The view ahead

Why does it matter whether some of these unusual animals fall into phyla of their own? The key reason is that every time we place a particular body plan, or unique morphology, onto the phylogenetic tree of animal life, it changes our perspective about the path of evolution. Consider the Xenoturbellida and Acoelomorpha. Animals in each of these groups have bilateral symmetry, but they lack a major central nerve cord in the midline of the body. This is, of course, in contrast to the situation in the majority of Bilateria – most ecdysozoans, lophotrochozoans, and deuterostomes have a main nerve cord. If the centralized nerve cord is a general feature of bilaterian animals, perhaps these two new phyla descended from very early branches in the evolutionary tree of animals? Did Xenoturbellida and Acoelomorpha, or perhaps just one of them, branch off before the divergence of Ecdysozoa, Lophotrochozoa, and Deuterostomia (but after Cnidaria)? If so, perhaps they provide us with a tantalizing glimpse of how the first bilaterally symmetrical animal bodies could have functioned, before the evolution of a main nerve cord for integrating information. Initial molecular analyses suggested this was indeed the case, at least for Acoelomorpha, although the conclusion was controversial. An alternative molecular study places Acoelomorpha and Xenoturbellida among the deuterostomes, alongside the echinoderms, hemichordates, and chordates. If this is correct, why do they not have a main nerve cord? Have they lost it in evolution, by spreading the nervous system around the body? Or are we wrong in our views about the common ancestor of Bilateria? These are important questions to address, but they hinge on pinning down exactly where Xenoturbellida and Acoelomorpha fit in the tree of life, something that has proved surprisingly difficult even with large amounts of molecular data.

This controversy brings us onto whether we should have confidence in the current phylogenetic tree of the Animal Kingdom. The 'new phylogeny' sees some early diverging non-bilaterian lineages (Porifera, Placozoa, Ctenophora, Cnidaria) separating from the branch leading to Bilateria, which in turn splits into three great superphyla: Ecdysozoa, Lophotrochozoa, and Deuterostomia. How sure can we be of this scenario? Hypotheses of evolutionary relationships have changed drastically over the past century, so might they change again? I predict they will not. Instead, I argue that it is time to have confidence in the 'new animal phylogeny', at least in broad outline. The phylogeny is based almost entirely on comparison of DNA sequences from genes found in all animals. And while the earliest molecular trees were built from one or a few genes, the basic framework has since been corroborated by massive analyses involving over a hundred genes per species. DNA sequence provides a mine of information on past history and, although not straightforward to analyse, it has provided the most robust, and the most internally consistent, data set ever applied to these problems. It is true that a few animals, such as Acoelomorpha, have not proved easy to place even using molecular data, but at least these methods leave them as unplaced or controversially placed, not squeezed into the most convenient place.

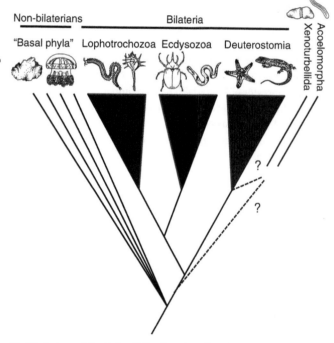

15. Phylogeny of the Animal Kingdom showing alternative hypotheses for the position of Xenoturbellida and Acoelomorpha

I believe we are at a time in the history of zoology when we have, for the first time, a robust evolutionary tree of animal diversity. We must remember, however, that this phylogenetic tree is just a starting point for biological investigation. A tree in itself does not provide understanding. What it provides is a framework that allows us to interpret biological data carefully and rigorously. Morphological studies, formerly used to build trees, now become more valuable than ever before as they can be interpreted in the light of an independent tree. Only with the robust framework of a phylogenetic tree can we compare anatomy, physiology, behaviour, ecology, and development between animal species in a meaningful way: a way that gives insight into the pattern and process of biological evolution.

Further reading

Chapter 1

L. W. Buss, *The Evolution of Individuality* (Princeton: Princeton University Press, 1987)

Chapter 2

A. L. Panchen, *Classification, Evolution and the Nature of Biology* (Cambridge: Cambridge University Press, 1992)

J. A. Valentine, *On the Origin of Phyla* (Chicago: University of Chicago Press, 2004)

Chapter 3

M. J. Telford and D. T. J. Littlewood (eds.), *Animal Evolution – Genomes, Fossils and Trees* (Oxford: Oxford University Press, 2009)

Chapter 4

R. Dawkins, *The Ancestor's Tale* (Boston: Houghton Mifflin, 2004)

Chapter 5

R. A. Raff, *The Shape of Life: Genes, Development, and the Evolution of Animal Form* (Chicago: University of Chicago Press, 1996)

S. B. Carroll, *Endless Forms Most Beautiful: The New Science of Evo Devo and the Making of the Animal Kingdom* (New York: W. W. Norton, 2005)

Chapter 6

R. B. Clark, *Dynamics in Metazoan Evolution: The Origin of the Coelom and Segments* (Oxford: Clarendon Press, 1964)

J. A. Pechenik, *Biology of the Invertebrates*, 3rd edn. (New York: McGraw-Hill, 2009)

Chapter 7

D. Grimaldi and M. Engel, *Evolution of the Insects* (Cambridge: Cambridge University Press, 2005)

Chapter 8

H. Gee, *Before the Backbone* (London: Chapman & Hall, 1996)

Chapter 9

J. A. Long, *The Rise of Fishes* (Baltimore: Johns Hopkins University Press, 2010)

Chapter 10

F. H. Pough, C. M. Janis, and J. B. Heiser, *Vertebrate Life*, 5th edn. (New Jersey: Prentice Hall, 1999)

Index

Expand your collection of
VERY SHORT INTRODUCTIONS